Cement-Based Composites

Cement-Based Composites

Advancements in Development and Characterization

Editors

Pawel Sikora
Sang-Yeop Chung

MDPI • Basel • Beijing • Wuhan • Barcelona • Belgrade • Manchester • Tokyo • Cluj • Tianjin

Editors
Pawel Sikora
Technische Universität Berlin
Germany
West Pomeranian University
of Technology Szczecin
Poland

Sang-Yeop Chung
Department of Civil and
Environmental Engineering,
Sejong University
Republic of Korea

Editorial Office
MDPI
St. Alban-Anlage 66
4052 Basel, Switzerland

This is a reprint of articles from the Special Issue published online in the open access journal *Crystals* (ISSN 2073-4352) (available at: https://www.mdpi.com/journal/crystals/special_issues/ Cement_Composites).

For citation purposes, cite each article independently as indicated on the article page online and as indicated below:

LastName, A.A.; LastName, B.B.; LastName, C.C. Article Title. *Journal Name* **Year**, *Article Number*, Page Range.

ISBN 978-3-03943-657-6 (Hbk)
ISBN 978-3-03943-658-3 (PDF)

Contents

About the Editors

Pawel Sikora is an Assistant Professor at the Faculty of Civil and Environmental Engineering of the West Pomeranian University of Technology, Szczecin. He received his Ph.D. in civil engineering from Warsaw University of Technology. He has vast teaching and research experience and is the co-author of numerous papers, mainly, in the field of modification of cementitious composites with nanomaterials and lightweight concretes. Currently, he is conducting his postdoctoral research stay at the Technical University of Berlin within the Marie-Skłodowska Curie Actions – Individual Fellowship (Horizon 2020) related to 3D printing of concrete.

Sang-Yeop Chung is working as an assistant professor in the Department of Civil and Environmental Engineering at Sejong University, Seoul, Korea. He attained his Ph.D. degree at Yonsei University in Korea and worked as a postdoctoral researcher at the same University. Prof. Chung also worked at Technische Universität Berlin (TU Berlin) in Germany until February 2019. He mainly studies the micro-scale material characterization and property evaluation of engineering materials (e.g., concrete, lightweight aggregate, metals). Throughout his academic career, Prof. Chung has received awards from the German Academic Exchange Service (DAAD) and the National Research Foundation of Korea (NRF) and has published many research papers in renowned international journals.

Editorial

Cement-Based Composites: Advancements in Development and Characterization

Pawel Sikora [1,2,*] and Sang-Yeop Chung [3,*]

[1] Building Materials and Construction Chemistry, Technische Universität Berlin, Gustav-Meyer-Allee 25, 13355 Berlin, Germany

[2] Faculty of Civil and Environmental Engineering, West Pomeranian University of Technology, Al. Piastow 50, 70-311 Szczecin, Poland

[3] Department of Civil and Environmental Engineering, Sejong University, 209 Neungdong-ro, Gwangjin-gu, Seoul 05006, Korea

* Correspondence: pawel.sikora@zut.edu.pl (P.S.); sychung@sejong.ac.kr (S.-Y.C.)

Received: 16 September 2020; Accepted: 17 September 2020; Published: 17 September 2020

Abstract: This Special Issue on "Cement-Based Composites: Advancements in Development and Characterization" presents the latest research and advances in the field of cement-based composites. This special issue covers a variety of experimental studies related to fibre-reinforced, photocatalytic, lightweight, and sustainable cement-based composites. Moreover, simulation studies are present in this special issue to provide the fundamental knowledge on designing and optimizing the properties of cementitious composites. The presented publications in this special issue show the most recent technology in the cement-based composite field.

Keywords: cement-based composite; experimental studies; numerical simulation; sustainability

---

1. Introduction

Concrete, a composite material composed of cement, water, aggregates, and often admixtures, is the most produced human-made material in the world. This material is an indispensable element of modern societies and is used in most of today's constructed engineering structures. Concrete structures need to satisfy specific characteristics in terms of mechanical performance and long-term durability so that they can be used without serious consideration of maintenance for many decades. Therefore, proper methods to produce advanced high-performance composites are actively required. Due to their composite nature, the choice of proper individual components and their interaction and compatibility plays a vital role in shaping the final properties of cement-based composites. In addition, there is a strong need to develop sustainable cementitious composites as well as their alternatives to tackle the challenge of increasing total anthropogenic carbon dioxide emissions. Therefore, numerical approaches to modeling and evaluating a material's characteristics and properties can also be used to accelerate the material's development.

This Special Issue on "Cement-Based Composites: Advancements in Development and Characterization" presents the latest research and advances in the field of cement-based composites.

2. Contents of this Special Issue

Marcalikova et al. [1] performed a study on the effects of two types of fibers (hooked and straight) on the quantitative and qualitative parameters of concrete. The fibers were hooked and straight. The influence of the fibers type and content on the mechanical properties in fiber-reinforced concrete was analyzed by functional dependence. Comprehensive evaluations of the mechanical properties of compressive strength, splitting tensile strength, bending tensile strength, and fracture energy were performed. Moreover, the resulting load-displacement diagrams and summary recommendations for the structural use and design of fiber-reinforced concrete structures were presented.

Smarzewski [2] conducted his research on local bond strength of short length specimens in high-performance concrete (HPC) and basalt fibre reinforced high-performance concrete (BFRHPC). As the main variables in this study, the basalt fibre volume content, concrete cover, bar diameter, and rib geometry were chosen. Moreover, additional factors were the directions of the casting and loading. For different ranges of BFRHPC strength, relationships for bond strength concerning the splitting tensile strength were obtained. The study showed that the bond strength increased with the splitting tensile strength and compressive strength of BFRHPC specimens with the 12 mm and 16 mm bar respectively. Moreover, the bond strength of BFRHPC was lower for the bar with the greater distances between the lugs on the bar.

In the study by Jiang et al. [3], a new mixture composition of steel fiber-reinforced MgO concrete (SFRMC) was proposed and evaluated to combine positive effects of both fiber-reinforced concrete and expansion concrete. The influence of steel fiber and MgO on the strength and chloride diffusion resistance of concrete was evaluated by splitting tensile test and chloride diffusion test. The results showed that the combined action of steel fiber and MgO reduced the porosity of concrete and the chloride diffusion coefficient, which could not be achieved by steel fiber and MgO separately. Moreover, mercury intrusion porosimeter (MIP) and scanning electron microscopy (SEM) confirmed that newly developed concretes exhibit lower porosities than reference concrete.

Cheng et al. [4] performed a study on the unconfined compression strength of polypropylene fiber-reinforced composite cemented clay. Three main factors, including polypropylene fiber content, composite cement content, and curing time on the unconfined compressive strength of fiber-reinforced cemented clay, were studied. The authors concluded that the incorporation of fibers can increase the compressive strength and residual strength of cement-reinforced clay as well as the corresponding axial strain when the stress peak is reached compared with cement-reinforced clay. The compressive strength of fiber-composite cement-reinforced marine clay increases with the increase of curing time and composite cement content. Moreover, differences in the failure mode in specimens were found: cement-reinforced clay is a brittle failure, while the failure mode of fiber-reinforced cemented clay is a plastic failure.

Lehner et al. [5] evaluated the chloride ion diffusion coefficient of self-compacting concrete with steel-fibre reinforcement. Data was compared with Ordinary Portland Cement (OPC). Three different procedures of diffusion coefficient calculation were presented namely: rapid chloride penetration test accelerated penetration tests with chloride, and surface measurement of electrical resistivity using Wenner probe. The resulting diffusion coefficients obtained by all methods are compared and evaluated regarding the basic mechanical properties of concrete mixtures. Meaningful relations between the testing methods and steel fibers content were established.

Zhong et al. [6] performed a study on visible light catalysis of graphite carbon nitride-silica composite material and its potential as a surface treatment of cement. Graphite carbon nitride-silica composite materials were synthesized by thermal polymerization using nanosilica and urea as raw materials. The effect of nanosilica content and specific surface area were investigated towards optimizing the material composition. The surface of cement-based materials was treated with graphite carbon nitride-silica composite materials by the one-sided immersion and brushing methods for the study of photocatalytic performance. By comparing the degradation effect of Rhodamine B, it was found that the painting method is more suitable for the surface treatment of cement. In addition, through the reaction of calcium hydroxide and graphite carbon nitride-silica composite materials, it was found that the combination of graphite carbon nitride-silica composite materials and cement is through C–S–H gel. Another approach towards the production of photocatalytic cementitious composites was proposed by Wang et al. [7]. $TiO_2@SiO_2$ core-shell nanocomposites with different coating thicknesses were synthesized by varying the experiment parameters. Authors reported that the introduction of SiO_2 coatings accelerated the rhodamine B removal to a certain extent, owing to its high surface area; however, more SiO_2 coatings decreased its photocatalytic efficiencies. The cement matrix treated with $TiO_2@SiO_2$ core-shell nanocomposites showed good photocatalytic efficiency and durability after harsh weathering processing. A reaction mechanism was revealed by the reaction of $TiO2@SiO_2$ nanocomposites with $Ca(OH)_2$.

Abd Elrahman et al. [8] performed a study towards the development of insulating light-weight cementitious composites in order to reduce the energy loss and consumption in buildings. Three different approaches towards incorporating air voids in cement pastes were proposed by introducing: aluminum powder, air-entraining agent, and hollow microspheres. A comprehensive evaluation of oven-dry density, compressive strength, porosity, water absorption, and thermal conductivity was performed. Moreover, X-ray micro-computed tomography (micro-CT) was adopted to investigate the microstructure of the air-entrained cement pastes. Meaningful relations between each type of air-entraining method were established. The experimental results obtained showed that specimens with an admixture of hollow microspheres can improve the compressive strength of cement composites compared to other air-entraining admixtures at the same density level. It was also confirmed that the incorporation of aluminum powder creates large voids, which have a negative effect on specimens' strength and absorption.

Hwang et al. [9] studied the effects of the aggregate surface conditions on the compressive strength of quick-converting track concrete (QTC). The compressive strength of QTC and interfacial fracture toughness (IFT) were investigated by changing the amount of fine abrasion dust particles (FADPs) on the aggregate surface from 0.00 to 0.15 wt% and the aggregate water saturation from 0 to 100%. Significant relations between the effects of aggregate water saturation on the compressive strength of the QTC and IFT were reported in correspondence to the amount of FADPs. As an outcome of this comprehensive research, fundamental knowledge of the importance of the aggregate surface conditions for the strength development of QTC was presented.

Iștoan et al. [10] performed a study on the production of lightweight mortars containing hemp shiv, volcanic rocks, and white cement. A complex study covering the evaluation of the chemical, acoustic, thermal, and mechanical properties as well as fire (heat) resistance were performed. Interesting relations between the constituents and effects on the various material on the selected properties were comprehensively evaluated. As an outcome of this study, sustainable cementitious mortar with increased indoor comfort performance was produced.

Ren et al. [11] performed a study towards understanding the underlying mechanism of inhibition of the alkali-carbonate reaction (ACR) using fly ash. Authors reported that when the alkali equivalent (equivalent Na_2Oeq) of the cement is 1.0%, the addition of 30% fly ash can significantly inhibit the expansion in low-reactivity aggregates, while for moderately reactive aggregates, the expansion rate can also be reduced by adding 30% of fly ash. The study showed that fly ash refines the pore structure of the cement paste, thus the alkali migration rate in the curing solution to the interior of the concrete microbars is reduced. As the content of fly ash increases, the concentrations of K^+ and $Na^{+'}$ and the pH value in the pore solution gradually decrease. This makes the ACR in the rocks slower, such that the cracks are reduced, and the expansion due to the ACR is inhibited.

In addition to the experimental studies, Kim et al. [12] proposed a numerical method for inversely estimating the spatial distribution characteristic of a material's elastic modulus using the measured value of the observation data and the distance between the measurement points. In this study, the structural factors with randomness are typically modeled as having a certain probability distribution (probability density function) and a probability characteristic (mean and standard deviation). To overcome the limitations of previous studies with uncertainties, Kim et al. propose a method to numerically define the spatial randomness of the material's elastic modulus and confirm factors such as response variability and response variance so that the material properties can be predicted using the proposed method.

The above-mentioned articles are of cement-based composites with different approaches, and these studies can contribute to the development of advanced cement-based materials significantly.

Author Contributions: Writing—original draft, P.S. and S.-Y.C.; Writing—review & editing, P.S. and S.-Y.C. All authors have read and agreed to the published version of the manuscript.

Funding: P.S. wishes to acknowledge that this special issue was launched thanks to his Postdoctoral research stay which has received funding from the European Union's Horizon 2020 research and innovation program under the Marie Skłodowska-Curie grant agreement No. 841592.

Acknowledgments: We thank all the authors who contributed to this special issue for preparing interesting papers as well as reviewers who shared their valuable time for reviewing these publications. P.S. is Supported by the Polish Foundation for Science.

Conflicts of Interest: The authors declare no conflict of interest.

References

1. Marcalikova, Z.; Čajka, R.; Bilek, V.; Bujdos, D.; Sucharda, O. Determination of Mechanical Characteristics for Fiber-Reinforced Concrete with Straight and Hooked Fibers. *Crystals* **2020**, *10*, 545. [CrossRef]

2. Smarzewski, P. Study of Bond Strength of Steel Bars in Basalt Fibre Reinforced High Performance Concrete. *Crystals* **2020**, *10*, 436. [CrossRef]

3. Jiang, F.; Deng, M.; Mo, L.; Wu, W. Influence of Combined Action of Steel Fiber and MgO on Chloride Diffusion Resistance of Concrete. *Crystals* **2020**, *10*, 338. [CrossRef]

4. Cheng, Q.; Zhang, J.; Zhou, N.; Guo, Y.; Pan, S. Experimental Study on Unconfined Compression Strength of Polypropylene Fiber Reinforced Composite Cemented Clay. *Crystals* **2020**, *10*, 247. [CrossRef]

5. Lehner, P.; Konečný, P.; Ponikiewski, T. Comparison of Material Properties of SCC Concrete with Steel Fibres Related to Ingress of Chlorides. *Crystals* **2020**, *10*, 220. [CrossRef]

6. Zhong, W.; Wang, D.; Jiang, C.; Lu, X.; Zhang, L.; Cheng, X. Study on Visible Light Catalysis of Graphite Carbon Nitride-Silica Composite Material and Its Surface Treatment of Cement. *Crystals* **2020**, *10*, 490. [CrossRef]

7. Wang, D.; Geng, Z.; Hou, P.; Yang, P.; Cheng, X.; Huang, S. Rhodamine B Removal of TiO_2@SiO_2 Core-Shell Nanocomposites Coated to Buildings. *Crystals* **2020**, *10*, 80. [CrossRef]

8. Elrahman, M.A.; El Madawy, M.E.; Chung, S.-Y.; Majer, S.; Youssf, O.; Sikora, P. An Investigation of the Mechanical and Physical Characteristics of Cement Paste Incorporating Different Air Entraining Agents using X-ray Micro-Computed Tomography. *Crystals* **2020**, *10*, 23. [CrossRef]

9. Hwang, R.; Lee, I.; Pyo, S.; Kim, D.J. Influence of the Aggregate Surface Conditions on the Strength of Quick-Converting Track Concrete. *Crystals* **2020**, *10*, 543. [CrossRef]

10. Istoan, R.; Tămaș-Gavrea, D.R.; Manea, D.L. Experimental Investigations on the Performances of Composite Building Materials Based on Industrial Crops and Volcanic Rocks. *Crystals* **2020**, *10*, 102. [CrossRef]

11. Ren, X.; Li, W.; Mao, Z.; Deng, M. Inhibition of the Alkali-Carbonate Reaction Using Fly Ash and the Underlying Mechanism. *Crystals* **2020**, *10*, 484. [CrossRef]

12. Kim, D.-Y.; Sikora, P.; Araszkiewicz, K.; Chung, S.-Y. Inverse Estimation Method of Material Randomness Using Observation. *Crystals* **2020**, *10*, 512. [CrossRef]

 *crystals*

Article

An Investigation of the Mechanical and Physical Characteristics of Cement Paste Incorporating Different Air Entraining Agents using X-ray Micro-Computed Tomography

Mohamed Abd Elrahman [1,2], Mohamed E. El Madawy [2], Sang-Yeop Chung [3,*], Stanisław Majer [4], Osama Youssf [2,5] and Pawel Sikora [1,4]

1   Building Materials and Construction Chemistry, Technische Universität Berlin, Gustav-Meyer-Allee 25, 13355 Berlin, Germany; abdelrahman@tu-berlin.de (M.A.E.); pawel.sikora@zut.edu.pl (P.S.)
2   Structural Engineering Department, Mansoura University, Elgomhouria St., Mansoura City 35516, Egypt; m_eltantawy@mans.edu.eg (M.E.E.M.); Osama.Youssf@mymail.unisa.edu.au (O.Y.)
3   Department of Civil and Environmental Engineering, Sejong University, 209 Neungdong-ro, Gwangjin-gu, Seoul 05006, Korea
4   Faculty of Civil Engineering and Architecture, West Pomeranian University of Technology, Szczecin, Al. Piastow 50, 70-311 Szczecin, Poland; stanislaw.majer@zut.edu.pl
5   School of Natural and Built Environments, University of South Australia, Mawson Lakes Blvd, Mawson Lakes SA 5095, Australia
*   Correspondence: sychung@sejong.ac.kr; Tel.: +82-2-6935-2471

Received: 13 December 2019; Accepted: 4 January 2020; Published: 6 January 2020

Abstract: Improving the thermal insulation properties of cement-based materials is the key to reducing energy loss and consumption in buildings. Lightweight cement-based composites can be used efficiently for this purpose, as a structural material with load bearing ability or as a non-structural one for thermal insulation. In this research, lightweight cement pastes containing fly ash and cement were prepared and tested. In these mixes, three different techniques for producing air voids inside the cement paste were used through the incorporation of aluminum powder (AL), air entraining agent (AA), and hollow microspheres (AS). Several experiments were carried out in order to examine the structural and physical characteristics of the cement composites, including dry density, compressive strength, porosity and absorption. A Hot Disk device was used to evaluate the thermal conductivity of different cement composites. In addition, X-ray micro-computed tomography (micro-CT) was adopted to investigate the microstructure of the air-entrained cement pastes and the spatial distribution of the voids inside pastes without destroying the specimens. The experimental results obtained showed that AS specimens with admixture of hollow microspheres can improve the compressive strength of cement composites compared to other air entraining admixtures at the same density level. It was also confirmed that the incorporation of aluminum powder creates large voids, which have a negative effect on specimens' strength and absorption.

Keywords: lightweight cement paste; air entraining agents; hollow microspheres; aluminum powder; pores; micro-CT; thermal insulation; compressive strength

---

1. Introduction

Nowadays, rapid population growth and development have led to a significant increase in energy demand. However, fossil fuels-which are the main energy resource-are limited and subject to being depleted in the next few decades. In addition, their use has had a harmful impact on the environment. They are responsible for climate change and environmental pollution; fossil fuels are the primary resource for CO_2 emission globally [1]. To cope with these challenges, it is imperative to make optimum use of the available energy resources and to reduce energy consumption as much as possible. In reference to thermal comport, buildings consume about 40% of global energy consumption and are consequently responsible for about one-third of global carbon emissions [2]. High efficiency energy utilization in buildings has therefore become a principle policy for many countries. Recently, there has been increasing interest in improving the thermal insulation of buildings in order to save energy and to reduce the environmental problem. In case of existing structures, this goal can be mostly achieved by insulating the already existing structures (e.g. masonry walls), which occurs mostly during the restoration period [3]. However, for a purpose of erecting the new structures several insulating materials were developed for decreasing the energy losses from buildings and to meet structural stability and strength requirements. Lightweight cement-based materials are one of the most ideal materials for such applications, because of its high thermal insulation efficiency as well as its reasonable mechanical characteristics [4–9].

Today, lightweight concrete is available in a wide range of densities and strengths which makes it suitable for various applications. The basic advantages of lightweight concrete are its low thermal conductivity, which makes it ideal for energy conservation, as well as its low density, which reduces the dead load of structures. Cellular concrete is a type of lightweight concrete composed of cement, water, entrained air and sometimes fine aggregate [10–12]. Generally, two methods have been developed for entraining air in concrete; physically, with the use of pre-formed foam (foamed concrete) and chemically, through the use of an air-entraining agent (aerated concrete). Cellular concrete can be designed to have a density of 200 to 1800 kg/m^3, depending on whether its applications are structural or non-structural [13]; though the density of cellular or lightweight concrete is relatively low, these materials have advantages to be used for both purposes. For example, structural lightweight concrete is largely used for building located in seismic area to reduce the mass subjected to seismic loads [14,15], and non-structural lightweight concrete is generally used for insulation and noise reduction [16,17].

Aerated concrete can be autoclaved (AAC) or non-autoclaved (NAAC) depending on the curing method used. In general, NAAC is known to be beneficial than AAC in the following aspects; NAAC is less expensive in production and shows better acoustic protection, fire safety, and better insulation than AAC [18–20]. On the basis of these advantages, between these approaches, the goal of the present research is to examine the properties of non-autoclaved aerated concrete, produced using different air-entraining agents (AEAs). NAAC is manufactured by creating macro porosity in a cement matrix, with the help of an expansive agent which reacts with the water and lime in the suspension [21]. This reaction liberates gas which expands the fresh concrete and generate pores, which reduce the weight and thermal conductivity of aerated concrete after hardening. The formed bubbles are nearly spherical [22] and have a diameter of about 10 μm to 1 mm [23] depending on the surfactant used. In addition to reducing the weight of concrete, air entraining agents are widely used to enhance concrete resistance to damage caused by freezing and thawing cycles. The type of air entraining agent has an important influence on the properties of aerated concrete [24,25]. Aluminum powder is normally used to obtain gas bubbles in concrete. In addition, various companies have developed different types of air entraining agents and foaming agents for normal and light-weight concretes.

To investigate the effect of different materials on aerated concrete, several researchers used various materials. Schackow et al. [26] used vermiculite and EPS to identify their effects on mechanical and thermal properties, and pre-saturated bentonite was adopted to produce ultra-lightweight concrete with numerous pores [27]. Super absorbent polymers (SAPs) was utilized as physical air entrainment in cement mortars [28]. Naratha et al. [29] have studied the effects of supplementary materials of the strength and thermal conductivity of non-autoclaved aerated concrete. Aluminum powder was added as a pore-forming agent in the amount of 0.2 wt.% of the binder content. A moderate-strength concrete, in the range of 13–23 MPa and with a density of about 1800 kg/m^3, was developed. They reported that aerated concrete incorporating fly ash and silica fume has considerable advantages and low energy consumption, compared to autoclaved aerated concrete. In relation to this, Ramamurthy and

Narayanan [30] have showed that fly ash plays a pivotal role in aerated concrete without deteriorating compressive strength. Aguilar et al. [31] have studied the microstructure of aerated cement pastes with fly ash, metakaolin and sepiolite additions; concluding that the addition of fly ash increases closed porosity and produces a higher amount of calcium silicate hydrates (C-S-H), as detected using differential thermal analysis (DTA) and scanning electron microscopy (SEM).

Several kinds of research have been carried out to investigate the influence of various parameters on the properties of aerated concrete. For instance, Narayanan and Ramamurthy [25] have reviewed the structure and properties of aerated concrete which is more homogeneous than normal concrete because of the absence of coarse aggregates. They classified the aerated concrete according to the pore formation method, the binder type and the curing method. They reported that curing and pore-formation methods are the main factors controlling the properties and microstructure of aerated concrete. On the other hand, Yang et al. [32] and Chen et al. [33] have claimed that the water to solid ratio is the most important factor in controlling the properties of aerated concrete. As the water to solid ratio increases, the density and (to a lesser extent) compressive strength decrease [34,35]. It has also been asserted that there is a linear relationship between aluminum powder content and concrete density, with density decreasing as aluminum content increases [25].

Lightweight concrete can be used for structural and non-structural purposes depending on its characteristics. EN 206 defines structural lightweight concrete as a material with dry density ranges between 800 and 2000 kg/m^3 with a compressive strength higher than 13 MPa (LC12/13). Cement industry is one of the main parts for CO_2 emissions worldwide. To produce sustainable cement-based materials and to reduce greenhouse gases, alternative cementitious materials need to be used to replace cement without harmful impacts on concrete properties. For this purpose, fly ash can be used as alternative material of cement with high replacement level. The research presented here aims to produce lightweight cement paste and evaluate its mechanical, physical and thermal properties, so as to provide the most effective material with good material properties. Hence, proposed material can be further applied as a base for lightweight concrete production. For this purpose, we produced a set of lightweight pastes with different additives, such as aluminum powder, an air entraining agent, and hollow microspheres. Their physical properties including density, absorption, compressive strength, and thermal conductivity were then evaluated and compared using sensitive measurement tools. To figure out the pore characteristics, which strongly affect material properties, X-ray micro-computed tomography (micro-CT) was adopted and utilized to visualize the pore distribution inside the specimens. Micro-CT is a nondestructive testing method that uses X-ray and can describe the inner structure of an object without damaging the specimen. For lightweight cementitious composites, micro-CT can be utilized to investigate the characteristics of microstructural features [36–40]. With the results obtained, the effectiveness of each additives in achieving better mechanical and thermal lightweight composites properties was compared.

2. Preparation of Specimens

2.1. Materials

Ordinary Portland cement CEM I 42.5 N provided by HeidelbergCement GmbH (Germany) and fly ash (EFA-Füller HP, Baumineral, Germany) were used in this investigation, conforming to EN 197-1 and EN 450-1, respectively. The physical properties and chemical composition of cement and fly ash were measured experimentally in the laboratory, as shown in Table 1. The cement: fly ash ratio was kept at 1:3 by mass, for all mixes. The water/binder ratio was constant for all the mixes (w/b = 0.50). The w/b ratio was selected to achieve a workable cement composite with slump flow ranges between 420–480 mm according to EN 13350-5.

Three methods were used to generate air voids inside the cement paste. The first was the chemical expansion method, using aluminum powder with 0.5, 1, 2, and 3 wt.% binder content. The second method involved adding a tenside-based air entraining agent, with the same ratios as in the case of the aluminum powder: 0.5, 1, 2, and 3 wt.%. The third method involved adding prefabricated hollow plastic air bubbles (microspheres), with 0.5, 1, 2, and 3 wt.% mass of binder. Hollow microspheres are materials with spherical voids inside a plastic shell, having a diameter of 25–60 mm (median: 35 mm) and a density of 0.2 g/cm^3 [40]. Each method of creating artificial voids produces a different volume of air inside the paste. The dosage of the materials used for producing the air was therefore kept constant to investigate their effects on paste porosity.

Table 1. Chemical composition and physical properties of the raw materials [wt.%].

Material	CaO	SiO$_2$	Al$_2$O$_3$	Fe$_2$O$_3$	MgO	Na$_2$O	K$_2$O	SO$_3$	Blaine Fineness [cm^2/g]	Density [g/cm^3]
CEM I 42.5 N	65.0	19.9	4.6	3.1	1.7	0.3	0.5	3.0	3500	3.1
Fly ash	3.1	49.2	27.6	7.6	2.1	0.9	5.0	0.7	2877	2.3

2.2. Mix Proportions and Specimen Preparation

In this investigation, a total of thirteen cement pastes were prepared and produced (Table 2). Specimens with different additives were denoted as AA (air entraining agent), AL (aluminum powder), and AS (hollow microspheres), with the numbers after the specimen names presenting the amount of air entraining agent. To produce the specimens, a standard mortar mixer with 5 liter capacity and two mixing speed (140 rpm and 285 rpm according to EN 196-3) was used for mixing the components. After weighing the proportions of each material, fly ash and cement were firstly dry mixed for 1 min at the low speed, to ensure better distribution and homogeneity of the mixture. The water and air entraining agents were then added within thirty seconds and further mixed for 1 min at low speed. Consequently, there was a minute break to remove the cement paste adhering to the sides and the bottom of the bowl and to place it in the middle of the bowl. Finally, the mixer was run at high speed for 2 min. The mixing procedures for all the mixes were carried out in the same manner. Steel molds of $40 \times 40 \times 160$ mm^3 were used for manufacturing cement pastes. After casting, the specimens were covered with a plastic sheet and kept in chamber with controlled temperature and relative humidity of 21 ± 1 °C and 99%, respectively. After 24 h of casting, the samples were demolded. They were then cured under water at a temperature of 20 ± 1 °C. At the age of testing, the prisms were cut to small cubes $40 \times 40 \times 40$ mm^3. The samples were surface dried prior to undertaking compressive strength tests. The first mix was designed without any air entraining agent, having been designated as the reference mix.

Table 2. Mix compositions of the used cement paste specimens.

Mix	Cement [g]	Fly Ash [g]	w/b Ratio	Air Entraining Agent [g]	Aluminum Powder [g]	Hollow Microspheres [g]
A0	600	1800	0.5	-	-	-
AA0.5	600	1800	0.5	12	-	-
AA1	600	1800	0.5	24	-	-
AA2	600	1800	0.5	48	-	-
AA3	600	1800	0.5	72	-	-
AL0.5	600	1800	0.5	-	12	-
AL1	600	1800	0.5	-	24	-
AL2	600	1800	0.5	-	48	-
AL3	600	1800	0.5	-	72	-
AS0.5	600	1800	0.5	-	-	12
AS1	600	1800	0.5	-	-	24
AS2	600	1800	0.5	-	-	48
AS3	600	1800	0.5	-	-	72

3. Measurements of Characteristic Properties

3.1. Material Properties

3.1.1. Material Density

Dry density (ρ_{dry}) was determined according to EN 12390-7. After curing under water for 28 days, the submerged mass was measured (under water) to determine the actual volume of the samples. The saturated samples were then dried in an oven at 105 °C, to a constant mass. From the dry mass and volume obtained, the following formula was implemented to calculate the dry density (Equation (1)):

$$\rho_{dry} = m_{dry}/V \tag{1}$$

where m_{dry} is the mass of the sample (kg) and V is the volume of the sample (m^3).

3.1.2. Compressive Strength

For the compressive strength determination, samples with dimensions of $40 \times 40 \times 40$ mm^3 were used. The surface of the samples with aluminum powder was not smooth or homogeneous; therefore, the samples were polished before testing to remove lumps and indentations and to make their sides flat. A modern digital crushing machine (Zwick Roell, Berlin, Germany) was used for performing the compressive strength tests. Six cubes of each mix were tested, with the mean value taken under consideration in this investigation.

3.1.3. Thermal Properties

The thermal conductivity of the cement paste specimens was evaluated experimentally using a Hot Disk device (Hot Disk AB, Göteborg, Sweden) which meets the requirements of ISO 22007-2 [41]. In this method, two samples from the same mix were used, with the sensor laid between the samples. Before measuring, the samples were polished to make them flat to ensure that no gaps between the sensor and the surfaces of the samples. Subsequently, the samples were dried in an oven at a temperature of 105 ± 5 °C, until reaching a constant mass so as to remove moisture, which has a critical influence on the thermal properties of cement-based materials. The measurements were repeated three times for each sample at different places, with the mean value considered. More details about thermal property measurement using the Hot Disk can be found in [40].

3.1.4. Effective Water Porosity

The effective water porosity of the aerated cement pastes, being the accessible porosity by water, was measured using the simple method of water displacement. In this method, the sample must be saturated under water for at least 24 h. After measuring the wet mass, the submerged mass under water was determined. The saturation samples were then dried in an oven at 105 ± 5 °C, until constant mass. From the information of the mass in both dry (m_{dry}) and saturated states (m_{sat}), and with the mass under water (m_{under}), the effective water porosity (P) can be calculated with the following equation (Equation (2)) [42]:

$$P(\%) = \frac{m_{sat} - m_{dry}}{m_{sat} - m_{under}} \times 100\% \tag{2}$$

3.1.5. Water Absorption

The water uptake capabilities of a partially immersed specimen can be measured according to ISO standard (15148). This method involves placing a dried specimen in a container of water, such that the immersion depth of the specimen is 5 mm. The primary force that pulls water up into the specimen is capillary suction. Increases in the specimen mass are determined at set time periods. Several measurements were taken over a period of up to 24 h, and a straight line was fitted to the plot of the increase in mass, versus the square root of time. From this relationship, it was possible to determine the absorption coefficient of the cement-based materials.

3.2. Microstructure Evaluation Using X-Ray Micro-Computed Tomography (Micro-CT)

X-ray micro-computed tomography (micro-CT), a nondestructive investigation method, was used to estimate the spatial distribution of pores within the cement-based materials. This technique allows a series of cross-sectional images and microstructures to be obtained. Detailed pore characteristics, such as pore size distribution and pore density contour, were also examined using micro-CT images and imaging tools [43]. More details about this method can be found in [39], with only a brief explanation of the CT imaging procedure presented in this paper. Figure 1 shows the imaging process used in this study. First, a region of interest (ROI) is selected from an original image for effective examination of the sample. The original and ROI images were 8-bit images, represented by a value ranging from 0 (black) to 255 (white) in grayscale. Each image was composed of 1000×1000 pixels with a 27.0 μm pixel size. A binary image for classifying solids and pores was then generated, using the Otsu method [44] and an appropriate threshold value. The 3D sample for describing the pore structure was generated by consequent stacking of a series of binary images. Using the 3D images of the specimens, material characteristics, such as porosity and the spatial distribution of pores, can be effectively examined.

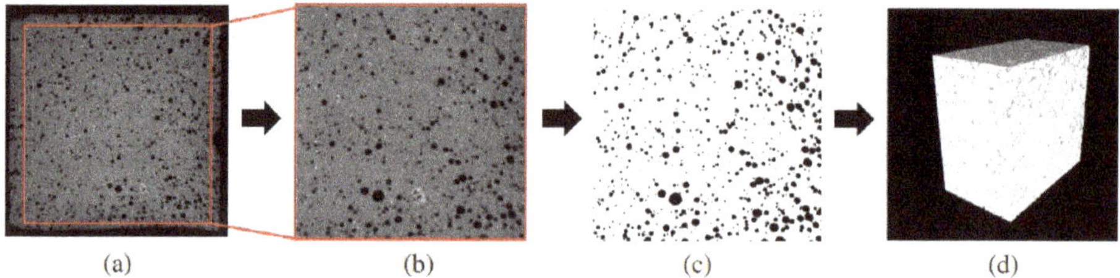

(a)   (b)   (c)   (d)

Figure 1. X-ray CT imaging process of the specimen with air-entraining admixture: (**a**) the original CT image, (**b**) region of interest (ROI), (**c**) binary image in 2D, (**d**) 3D binary image (Note: in the binary images, the white represents the solid phase, and the black regions are pores inside the specimen).

4. Results and Discussions

4.1. Material Properties of the Specimens

In this section, several properties of specimens with different additives and dosages are presented and discussed. The methods in Section 3.1 were adopted for each property.

4.1.1. Density

Figure 2 shows the experimental results of the dry density of cement composites. It can be clearly seen that in all cases, the density decreased as the amount of AA increased. Both AA and AS specimens showed similar effects on the density of the cement pastes; falling from 1.34 t/m^3 (reference - A0) to 1.06 t/m^3 and 1.04 t/m^3. for mixes with 3 wt.% of AA and AS, respectively. However, the aluminum powder was much more effective than either AA or AS in reducing density. Using 0.5% AL reduced it from 1.34 to 0.85 t/m^3. However, with a higher addition of AL, up to 3 wt.%, the density decreased only slightly more, to 0.75 t/m^3, as can be seen for the results of the AL3 specimen.

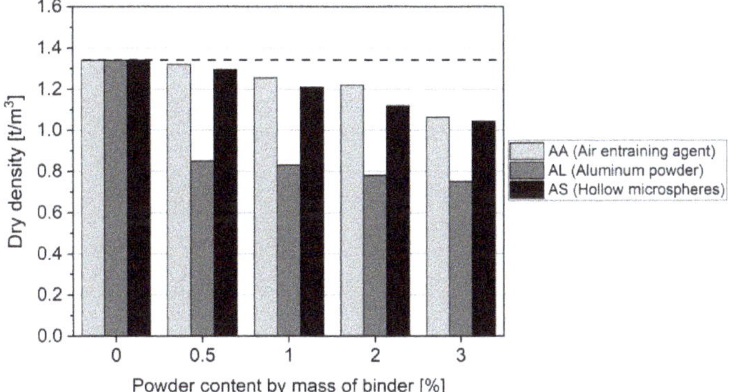

Figure 2. Dry density values of the specimens with different additives.

4.1.2. Compressive Strength

Compressive strength of concrete depends mainly on its major components; aggregate, cement paste, and transition zone in between. This study focuses on the properties of cement matrix which affect mechanical properties of cement-based materials significantly. Compressive strength of cement pastes was measured as an indicator of mechanical properties after curing for 28 days. Figure 3 shows the experimental results of the compression test on different pastes. In this figure, it is clearly presented that the incorporation of air voids into cement composites significantly decreases the compressive strength depending on the dosage and type of air entraining method. The AL specimens with aluminum powder show the lowest strength, which dropped from 30 MPa for reference mix without aluminum powder to 3.2 and 1.7 MPa for pastes with 0.5 and 3 wt.% aluminum powder, respectively. The same trend of the strength reduction with inclusion of air voids can be found in the case of the AA mixes, although the amount of the change is less. With the use of 3 wt.% of AA, the strength decreased to 18 MPa which represents about 60% of the reference mix. The influence of hollow microspheres (AS) on strength reduction is similar to that of the AA mixes. With addition of 3 wt.% of hollow microspheres, the compressive strength decreased to 15 MPa.

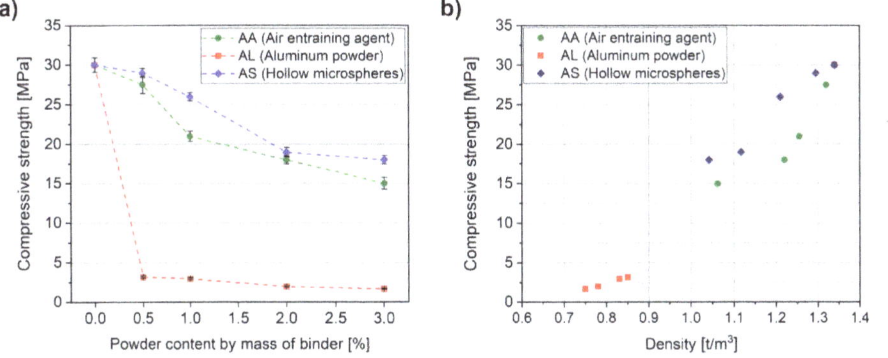

Figure 3. Compressive strength of the specimens with different additives: (**a**) compressive strength vs. powder content; (**b**) compressive strength vs. density.

Moreover, it is obvious from the results that 0.5 wt.% of aluminum powder reduced the strength significantly. This indicates that aluminum powder is more effective in producing air voids even with small dosage which reduces the density as well as the compressive strength. The influences of the other methods, AA and AS, on density and strength are not significant than that of aluminum powder even with higher dosages, as can be seen in Figure 3a. The relationship between the strength and dry density of cement pastes is presented in Figure 3b. The experimental results confirm that dry density has clear relationship with the compressive strength of cement pastes. The strength deterioration in the samples is directly attributable to increases in porosity volume and reductions in the solid structure volume. The strength retrogression was more pronounced in the case of aluminum powder, which was not only related to an increase in porosity volume, but also to the large size of the pores created, as detected by micro-CT measurements, which are presented below.

4.1.3. Effective Water Porosity

Water was used to evaluate cement paste porosity experimentally, using the water displacement method. Figure 4 illustrates the experimental results of total porosity at an age of 28 days. It can be seen that the reference mix (A0) had the lowest porosity of about 38 vol.%. For the other mixes, porosity increased with air content. With 3 wt.% AA, it increased to 50.1 vol.%; however, with 3 wt.% AS, it reached 51.7 vol.%. The increase in porosity was much higher in the mixes with aluminum powder. It increased from 38 vol.% without aluminum powder to 65 % with 3 wt.%. It is clear in the figure that the addition of 0.5 wt.% of aluminum powder increased the porosity to 60 vol.% which was more effective than 3 wt.% of AA or AS, where the porosity was about 50.1 vol.% and 51.7 vol.%, respectively. In addition, the experimental results indicate that higher dosages of aluminum powder do not significantly increase porosity in comparison to lower dosages; an increase in aluminum powder dosage from 0.5 wt.% to 3 wt.%, increased porosity from 60 to 65 vol.%. It can be seen from Figure 4 that increasing AA and AS dosages increased porosity gradually. However, in the case of AL, porosity increased primarily at the beginning, with the rate of increase obviously decreasing later with the increment in the AL content.

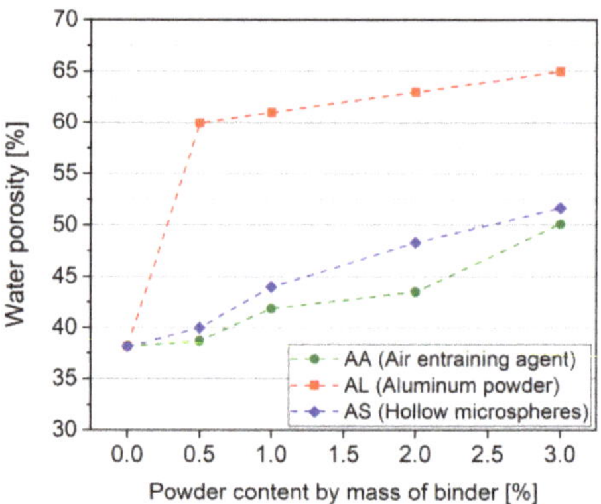

Figure 4. Water porosities of the specimens with different additives.

4.1.4. Water Absorption

Figure 5 presents the water absorption coefficient values of the different cement pastes. Mix A0, without entrained air, had the lowest water absorption, of about 0.61844 kg/(m^2·h$^{0.5}$). However, mix AA3 with 3 wt.% AA had the highest water absorption coefficient, of about 2.10211 kg/(m^2·h$^{0.5}$). In contrast to the porosity results, mixes with the aluminum powder had a lower absorption than mixes with the AA. For mixes with the hollow microspheres (AS), the absorption was higher than in the control mix, though a marginal increase in absorption was observed with an increase in AS dosage. Compared to AA mixes, those with the AS had a lower absorption coefficient. The capillary suction force controls absorption rate and quantity, in the method used to measure water absorption (water uptake). As a result, pores in the capillary range are more effective in increasing the water absorption coefficient. In the case of aluminum powder, most of the pores were large in size, and as a result, the water absorption of these mixes was smaller than in the other mixes, where most of the pores were in the size range of the capillary pores. It is clear from the results that, the water absorption of composites made with hollow microspheres is increasing up to 0.5 wt.%. With increment of AS content no significant increase in the absorption can be detected. This can be attributed to the nature of microspheres which are closed pores with weak shell and cannot create a continuous passage for fluids to transfer. Most of these pores do not participate in increasing the water absorption as can be reflected from the experimental results and this can be considered as an advantage of this material compared to other air entraining materials.

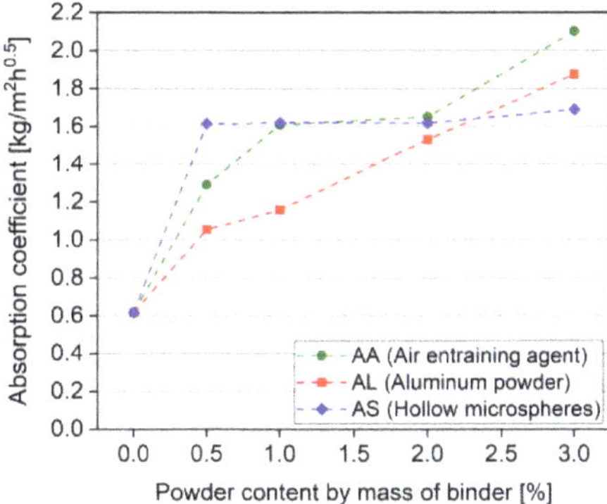

Figure 5. Water absorption of the specimens with different additives.

4.1.5. Thermal Conductivity

Figure 6 presents the experimental thermal conductivity results of the different cement paste samples. The thermal conductivity of the reference mix (A0) was about 0.347 W/(m·K). The thermal conductivity decreased as the AA content increased, reaching a value of 0.29 W/(m·K) at 3 wt.%. The experimental results for the mixes with hollow microspheres (AS) were similar to those with AA; conductivity fell from 0.32 W/(m·K) with 0.5 wt.% AS to 0.23 W/(m·K) with 3 wt.%. Aluminum powder seems to have a remarkable influence on thermal conductivity; with the addition of 0.5 wt.%, thermal conductivity dropped to 0.18 W/(m·K), while an addition of 3 wt.% aluminum powder reduced thermal conductivity to 0.12 W/(m·K).

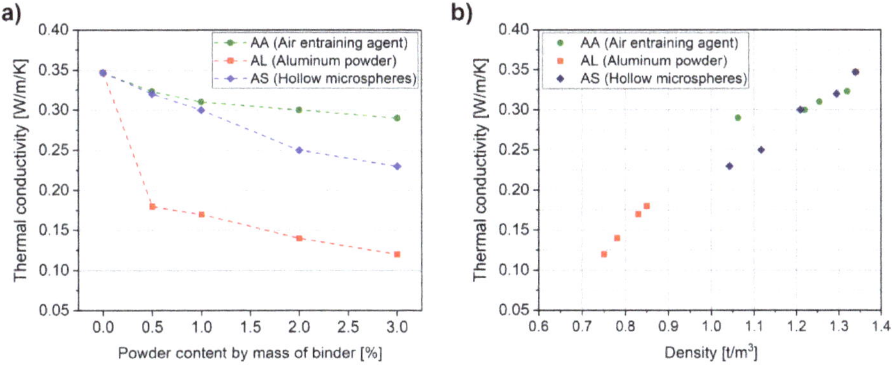

Figure 6. Thermal conductivity of the specimens with different additives: (**a**) thermal conductivity vs. powder content; (**b**) thermal conductivity vs. density.

Thermal conductivity is a material property related directly to dry density, as well as to the moisture content of the material. In this experiment, all of the samples were dried to a constant mass; as a result, the main factor influencing thermal conductivity was dry density [10]. Mixes with aluminum powder had the lowest thermal conductivity, due to reduced density as compared to other mixes, as can be seen in Figures 2 and 6. Heat transfer in any materials depends on its components: solid structure and voids. With decreasing material density, pore volume increases, and the heat transfer decreases; this influence can be clearly detected from the results of thermal conductivity of cement pastes with different densities.

4.2. Pore Characteristics Using X-Ray CT

The inner structures of the cement pastes specimens with different admixtures were examined using X-ray CT images. Figure 7a shows the binarized 3D micro-CT images of A0, AA3, AL3, and AS3 specimens, which were obtained without damaging the specimens. In the binary figures, the white region represents the solid part of the specimen, while the black region inside the specimen presents the pores. From this figure, the pore structure inside each specimen can be easily identified. In Figure 7a, AA and AS specimens with air entraining agent and hollow microspheres solid, respectively, have a relatively high porosity as compared to the reference (A0) specimen, with their spatial distribution seemingly relatively uniform. However, the AL specimen with aluminum powder had distinctively larger pores and porosity than the other specimens; it denotes that aluminum powder had a more significant effect on securing the pores inside the specimens, than did the other admixtures.

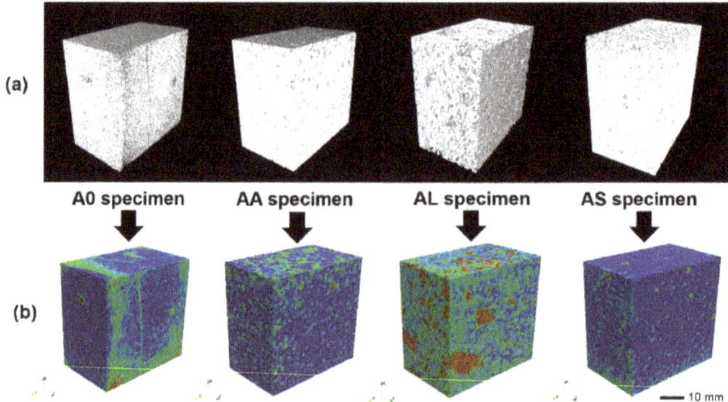

Figure 7. Target specimens (A0, AA3, AL3, AS3) and their 3D binarized CT images (solid structures): (a) 3D binarized CT images (solid structures) of A0, AA3, AL3, and AS3 specimens (Note: in the binary images, the white denotes solid phase, and the black parts represent the pores of each specimen); (b) description of pore densities of the specimens (Note: the color in each figure denotes pore density values between 0 (blue) to 1 (red)).

For a more detailed investigation of the pore structures using the 3D images, porosity, spatial pore density, and pore size distribution were also examined qualitatively and quantitatively. The porosity of each specimen was calculated based on the number of pore voxels in the 3D images; the obtained total porosities of the specimens were 14.5 vol.% (A0), 22.3 vol.% (AA3), 35.1 vol.% (AL3), and 19.9 vol.% (AS3). The porosity computed from the micro-CT measurement was much smaller than that of the water porosity because of the resolution limit of the used images. Since pores even in the order of a few micrometers or even less can be detected using the experimental approach, micro-CT porosity only concerns an analysis of pores larger than the image resolution (27.0 μm); the difference thus results from the minimum measurable pore size of each method. Despite their limitations, micro-CT images can be used to visualize pore structures both quantitatively and qualitatively. From the porosities obtained, the effects of each admixture on material porosity can be examined. The AL3 specimen shows the largest porosity which causes low thermal conductivity and mechanical strength. In the same manner, the AS3 specimen with the lowest porosity among the cases shows the largest compressive strength in Figure 3; these results confirmed that the porosity data obtained from micro-CT presents consist result with the measured properties of the specimens.

The spatial distribution of the pores in the specimens have been clearly described in Figure 7b, where the pore density of each specimen has been represented in visual form. To calculate pore density, the local porosity of the specimen was calculated for the whole specimen, after which computed values were assigned for every voxel of the specimen. For example, in the figure for the AL3 specimen, the red region indicates a region with a high local porosity, while the blue one represents a region with low local porosity. From these contours, the spatial distribution of pores in each specimen can be identified. In general, the pore distribution in the AL3 specimen can be considered as relatively uniform, showing an anisotropic trend, because the larger pore density values are distributed in a specific region. A similar phenomenon has been observed by Shabbar et al. [45] where aerated concretes containing higher aluminum powder contents (less than 0.5 wt.%) exhibited more anisotropic voids. From the results, both general trend of pore clustering as well as the effect of air entraining agents on the pore characteristics of the materials.

Figure 8 shows the pore size distribution in each specimen. From this figure, the relative pore sizes of the specimens can be compared, with all the specimens with air-entraining admixtures containing higher pore volume as well as pore sizes, particularly in the case of the AL3 specimen. It can therefore be confirmed that the pore distribution in the AL3 specimen was anisotropic, meaning that the use of aluminum powder in cement-based materials can induce a relatively large and anisotropic pore distribution. In addition, the data obtained shows consistent results in regard to other material properties. For example, the general trends regarding compressive strength and thermal conductivity, as shown in Figures 3 and 6, are similar to the pore size distribution and porosity results in Figures 7 and 8; both properties decreased as the amount of air-entraining agent or porosity increased. This result demonstrates that the use of micro-CT can be successfully utilized to describe the correlation between the material properties and the pore characteristics of cement-based materials.

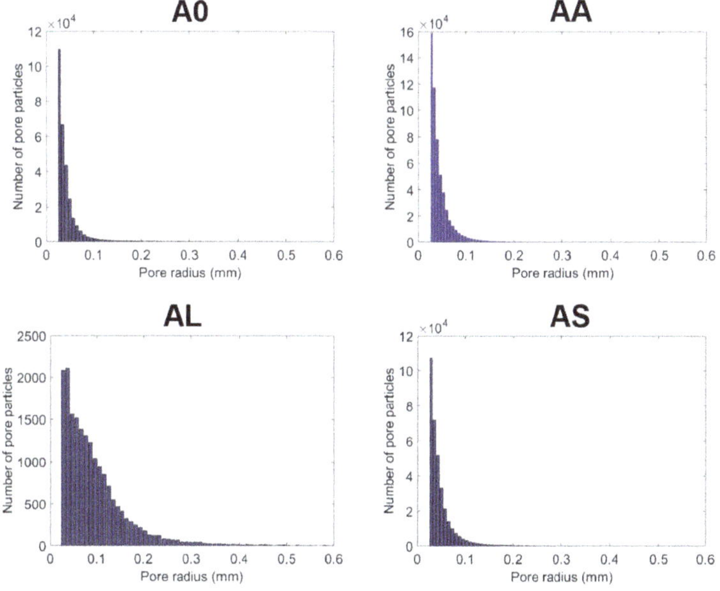

Figure 8. Pore size distribution of A0, AA3, AL3, and AS3 specimens.

Physical and chemical characteristics of fine materials affect the performance of air entraining agent (AA) as well as aluminum powder (AL) significantly, while their influence on hollow microsphere (AS) is very limited. The performance of AA and AL depends on the pH value of the pore solution and on the presence of other supplementary materials. The presence of fly ash influences the performance of AA negatively because the carbon particles in the fly ash absorb the agent and reduce their reactivity; consequently, higher dosages are needed. In addition, the carbon particles destabilize entrained air bubbles and change its volume in fresh status of cement-based materials [22]. In the case of aluminum powder, the alkali contents and cement fineness affect its performance significantly [46]. Other parameters, such as temperature, pH of the pore solution, and the consistency of the paste, influence the reactivity of aluminum powder [47]. In contrast, hollow microspheres are more stable and do not rely on the chemical composition of the used materials strongly. Only over-mixing for long period can destroy the voids and change their distribution. To avoid these problems in this investigation, cement and fly ash with a constant proportion of 1:3 was used for all mixes so that no influences can be occurred due to chemical composition. In addition, the mixing sequence and period were the kept the same for all mixes.

5. Conclusions

In this investigation, three different materials: aluminum powder (AL), air entraining agent (AA) and hollow microspheres (AS) were adopted to create pores inside cement pastes. The same amounts of each material were used to produce lightweight pastes with pores, with the effects of their dosage on material properties and characteristics investigated. The cement pastes' physical and mechanical properties were measured experimentally. In addition, paste microstructures were characterized using micro-CT images. The remarks below serve as a conclusion:

- An addition of aluminum powder increases air void contents more than other air-induced materials even with lower dosage. The addition of 0.5 wt.% of aluminum powder significantly reduces dry density, while higher dosages have negligible additional effects. The dry density of the specimens with air entraining agent and hollow microspheres are similar and reduced gradually with increasing the dosages.
- Dry density and compressive strength of aluminum powder pastes are much lower than both air entraining agent and hollow microspheres.
- Thermal conductivity is related directly to material density as reflected by the experimental results. Aluminum powder mixes had the lowest thermal conductivity, as compared to air entraining agent and hollow microspheres, with aluminum powder being the most effective for the purposes of insulation.
- Mixes with hollow microspheres showed lower thermal conductivity than mixes with air entraining agent due to different in density as well as pore size distribution.
- Among the used air entraining additives, the use of hollow microspheres is the most beneficial to achieve higher compressive strength. In contrast, aluminum powder can produce highly porous and large pores, as according to the micro-CT measurements. In particular, the use of aluminum powder can generate anisotropic pore clusters, which can cause a reduction in mechanical properties, while improving insulation.

As the results of this study have shown, each air-entraining agent has both advantages and disadvantages depending on thermal, mechanical, and absorption characteristics. The appropriate additive can therefore be selected according to the characteristics of the material at hand. In addition to the results of this study, the systematic investigation tools used here can be utilized in the further development of advanced lightweight cement-based materials as well as special building materials with particular objectives. In addition, the investigation of specimens with the same porosity but different air entraining agents can be conducted as a further study.

Author Contributions: Conceptualization, S.-Y.C. and M.A.E.; methodology, M.A.E. and M.E.E.M.; software, S.-Y.C.; validation, M.A.E., S-Y.C. and M.E.E.M.; formal analysis, M.A.E., S.-Y.C., and P.S.; investigation, M.A.E., M.E.E.M., S.M. and O.Y.; resources, P.S. and S.M.; data curation, S.-Y.C., M.A.E. and P.S.; writing—original draft preparation, M.A.E., S.-Y.C. and M.E.E.M.; writing—review and editing, S.-Y.C., M.A.E., P.S., O.Y. and S.M.; visualization, P.S. and S.-Y.C.; supervision, M.A.E. and M.E.E.M.; project administration, P.S. and M.A.E.; funding acquisition, P.S. and M.E.E.M. All authors have read and agreed to the published version of the manuscript.

Funding: This project has received funding from the European Union's Horizon 2020 research and innovation program under the Marie Skłodowska-Curie grant agreement No. 841592.

Acknowledgments: This research is supported by the German Egyptian Mobility Program for Scientific Exchange and Excellence Development (GE-SEED). P.S. is supported by the Foundation for Polish Science (FNP). The authors would like to express their appreciation to Paul H. Kamm (Helmholtz Centre Berlin) for his assistance in micro-CT imaging.

Conflicts of Interest: The authors declare no conflict of interest.

References

1. IEA. *CO_2 Emissions from Fuel Combustion Highlights*; International Energy Agency: Paris, France, 2016.
2. Tyagi, V.V.; Pandey, A.K.; Buddhi, D.; Kothari, R. Thermal performance assessment of encapsulated PCM based thermal management system to reduce peak energy demand in buildings. *Energy Build.* **2016**, *117*, 44–52. [CrossRef]
3. Foraboschi, P. Effectiveness of novel methods to increase the FRP-masonry bond capacity. *Comp. Part B Eng.* **2016**, *107*, 214–232. [CrossRef]
4. Chung, S.-Y.; Stephan, D.; Abd Elrahman, M.; Han, T.-S. Effects of anisotropic voids on thermal properties of insulating media investigated using 3D printed samples. *Constr. Build. Mater.* **2016**, *111*, 529–542. [CrossRef]
5. Chung, S.-Y.; Abd Elrahman, M.; Stephan, D.; Kamm, P.H. The influence of different concrete additions on the properties of lightweight concrete evaluated using experimental and numerical approaches. *Constr. Build. Mater.* **2018**, *189*, 314–322. [CrossRef]
6. Garbalińska, H.; Strzałkowski, J. Thermal and strength properties of lightweight concretes with variable porosity structures. *J. Mater. Civ. Eng.* **2018**, *30*, 04018326. [CrossRef]
7. Sadowski, Ł.; Popek, M.; Czarnecki, S.; Mathia, T.G. Morphogenesis in solidification phases of lightweight concrete surface at early ages. *Constr. Build. Mater.* **2017**, *148*, 96–103. [CrossRef]
8. Kurpinska, M.; Kułak, L. Predicting Performance of Lightweight Concrete with Granulated Expanded Glass and Ash Aggregate by Means of Using Artificial Neural Networks. *Materials* **2019**, *12*, 2002. [CrossRef]
9. Foraboschi, P. Masonry does not limit itself to only one structural material: Interlocked masonry versus cohesive masonry. *J. Build. Eng.* **2019**, *26*, 100831. [CrossRef]
10. Dorey, R.; Yeomans, J.; Smith, P. Effect of pore clustering on the mechanical properties of ceramics. *J. Eur. Ceram. Soc.* **2002**, *22*, 403–409. [CrossRef]
11. Amran, Y.H.; Farzadnia, N.; Ali, A.A. Properties and applications of foamed concrete; a review. *Constr. Build. Mater.* **2015**, *101*, 990–1005. [CrossRef]
12. Jaing, J.; Lu, Z.; Niu, Y.; Li, J.; Zhang, Y. Study on the preparation and properties of high-porosity foamed concretes based on ordinary portland cement. *Mater. Des.* **2016**, *92*, 949–959. [CrossRef]
13. Chung, S.-Y.; Lehmann, C.; Abd Elrahman, M.; Stephan, D. Pore characteristics and their effects on the material properties of foamed concrete evaluated using micro-CT images and numerical approaches. *Appl. Sci.* **2017**, *7*, 550. [CrossRef]
14. Foraboschi, P. Bending load-carrying capacity of reinforced concrete beams subjected to premature failure. *Materials* **2019**, *12*, 3085. [CrossRef] [PubMed]
15. Foraboschi, P. Structural layout that takes full advantage of the capabilities and opportunities afforded by two-way RC floors, coupled with the selection of the best technique, to avoid serviceability failures. *Eng. Fail. Anal.* **2016**, *70*, 387–418. [CrossRef]
16. Abd Elrahman, M.; Chung, S.-Y.; Sikora, P.; Rucinska, T.; Stephan, D. Influence of nanosilica on mechanical properties, sorptivity, and microstructure of lightweight concrete. *Materials* **2019**, *12*, 3078. [CrossRef] [PubMed]

17. Chung, S.-Y.; Kim, J.-S.; Stephan, D.; Han, T.-S. Overview of the use of micro-computed tomography (micro-CT) to investigate the relation between the material characteristics and properties of cement-based materials. *Constr. Build. Mater.* **2019**, *229*, 116843. [CrossRef]

18. Vesova, L.M. Disperse reinforcing Role in Producing Non-autoclaved Cellular Foam Concrete. *Procedia Eng.* **2016**, *150*, 1587–1590. [CrossRef]

19. Esmaily, H.; Nuranian, H. Non-autoclaved high strength cellular concrete from alkali activated slag. *Constr. Build. Mater.* **2012**, *266*, 200–206. [CrossRef]

20. Wang, C.-Q.; Lin, X.-Y.; Wang, D.; He, M.; Zhang, S.-L. Utilization of oil-based drilling cuttings pyrolysis residues of shale gas for the preparation of non-autoclaved aerated concrete. *Constr. Build. Mater.* **2018**, *162*, 359–368. [CrossRef]

21. Cabrillac, R.; Fiorio, B.; Beaucour, A.; Dumontet, H.; Ortola, S. Experimental study of the mechanical anisotropy of aerated concretes and of the adjustment parameters of the introduced porosity. *Constr. Build. Mater.* **2016**, *20*, 286–295. [CrossRef]

22. Neville, A.M. *Properties of Concrete*; Wiley: Chichester, UK, 2012; pp. 10–24.

23. Ley, M.T.; Chancey, R.; Juenger, M.; Folliard, K. The physical and chemical characteristics of the shell of air-entrained bubbles in cement paste. *Cem. Concr. Res.* **2009**, *39*, 417–425. [CrossRef]

24. Hilal, A.A.; Thom, N.H.; Dawson, A.R. On void structure and strength of foamed concrete made without/with additives. *Constr. Build. Mater.* **2015**, *85*, 157–164. [CrossRef]

25. Narayanan, N.; Ramamurthy, K. Structure and properties of aerated concrete: A review. *Cem. Concr. Compos.* **2000**, *22*, 321–329. [CrossRef]

26. Schackow, A.; Effting, C.; Folgueras, M.V.; Guths, S.; Mendes, G.A. Mechanical and thermal properties of lightweight concretes with vermiculite and EPS using air-entraining agent. *Constr. Build. Mater.* **2014**, *57*, 190–197. [CrossRef]

27. Xie, Y.; Li, J.; Lu, Z.; Jiang, J.; Niu, Y. Preparation and properties of ultra-lightweight EPS concrete based on pre-saturated bentonite. *Constr. Build. Mater.* **2019**, *195*, 505–514. [CrossRef]

28. Riyzai, S.; Kevern, J.T.; Mulheron, M.M. Super absorbent polymers (SAPs) as physical air entrainment in cement mortars. *Constr. Build. Mater.* **2017**, *147*, 669–676. [CrossRef]

29. Naratha, C.; Thongsanitgarn, P.; Chaipanich, A. Thermogravimetry analysis, compressive strength and thermal conductivity tests of non-autoclaved aerated Portland cement-fly ash-silica fume concrete. *J. Therm. Anal. Calorim.* **2015**, *122*, 11–20. [CrossRef]

30. Ramamurthy, K.; Narayanan, N. Factors influencing the density and compressive strength of aerated concrete. *Mag. Concr. Res.* **2000**, *52*, 163–168. [CrossRef]

31. Aguilar, A.S.; Melo, J.P.; Olivares, F.H. Microstructural analysis of aerated cement pastes with fly ash, metakaolin and Sepiolite additions. *Constr. Build. Mater.* **2013**, *47*, 282–292. [CrossRef]

32. Yang, L.; Yan, Y.; Hu, Z. Utilization of phosphogypsum for the preparation of non-autoclaved aerated concrete. *Constr. Build. Mater.* **2013**, *44*, 600–606. [CrossRef]

33. Chen, X.; Yan, Y.; Liu, Y.; Hu, Z. Utilization of circulating fluidized bed fly ash for the preparation of foam concrete. *Constr. Build. Mater.* **2014**, *54*, 137–146. [CrossRef]

34. Kearsley, E.P.; Wainwright, P.J. The effect of porosity on the strength of foamed concrete. *Cem. Concr. Res.* **2002**, *18*, 390–398. [CrossRef]

35. Korat, L.; Ducman, V.; Legat, A.; Mirtic, B. Characterisation of the pore-forming process in lightweight aggregate based on silica sludge by means of X-ray micro-tomography (micro-CT) and mercury intrusion porosimetry (MIP). *Cem. Concr. Res.* **2013**, *39*, 6997–7005. [CrossRef]

36. Gallucci, E.; Scrivener, K.; Groso, A.; Stampanoni, M.; Margaritondo, G. 3D experimental investigation of the microstructure of cement pastes using synchrotron X-ray microtomography. *Cem. Concr. Res.* **2007**, *37*, 360–368. [CrossRef]

37. Chung, S.-Y.; Han, T.-S.; Yun, T.S.; Yeom, K.S. Evaluation of the anisotropy of the void distribution and the stiffness of lightweight aggregates using CT imaging. *Constr. Build. Mater.* **2013**, *48*, 998–1008. [CrossRef]

38. Patterson, B.M.; Escobedo-Diaz, J.P.; Dennis-Koller, D.; Cerreta, E. Dimensional quantification of embedded voids or objects in three dimensions using X-ray tomograph. *Microsc. Microanal.* **2012**, *18*, 390–398. [CrossRef]

39. Chung, S.-Y.; Elrahman, M.A.; Kim, J.-S.; Han, T.-S.; Stephan, D.; Sikora, P. Comparison of lightweight aggregate and foamed concrete with the same density level using image-based characterizations. *Constr. Build. Mater.* **2019**, *211*, 988–999. [CrossRef]

40. Chung, S.-Y.; Abd Elrahman, M.; Stephan, D.; Kamm, P.H. Investigation of characteristics and responses of insulating cement paste specimens with Aer solids using X-ray micro-computed tomography. *Constr. Build. Mater.* **2016**, *118*, 204–215. [CrossRef]

41. ISO 22007-2:2015. Plastics–Determination of Thermal Conductivity and Thermal Diffusivity–Part 2: Transient Plane Heat Source (Hot Disk) Method. Available online: https://www.iso.org/standard/61190.html (accessed on 13 December 2019).

42. Abd Elrahman, M.; Hillemeier, B. Combined effect of fine fly ash and packing density on the properties of high performance concrete: An experimental approach. *Constr. Build. Mater.* **2014**, *58*, 225–233. [CrossRef]

43. *MATLAB Version 2018a*; The MathWorks Inc.: Natick, MA, USA, 2018.

44. Otsu, N. A threshold selection method from gray-level histograms. *IEEE Trans. Syst. Man Cybern.* **1979**, *9*, 62–66. [CrossRef]

45. Shabbar, R.; Nedwell, P.; Wu, Z. Mechanical properties of lightweight aerated concrete with different aluminum powder content. *MATEC Web of Conf.* **2017**, *120*, 02010. [CrossRef]

46. Ramaschndran, V.S. *Concrete Admixtures Handbook*; William Andrew: New York, NY, USA, 1996.

47. Mindess, S.; Young, J.F.; Darwin, D. *Concrete*; Prentice Hall: New York, NY, USA, 2002.

 *crystals*

Article

Rhodamine B Removal of TiO$_2$@SiO$_2$ Core-Shell Nanocomposites Coated to Buildings

Dan Wang [1], Zhi Geng [1], Pengkun Hou [1], Ping Yang [2], Xin Cheng [1,*] and Shifeng Huang [1,*]

[1] Shandong Provincial Key Laboratory of Preparation and Measurement of Building Materials, University of Jinan, Jinan 250022, Shandong, China; mse_wangd@ujn.edu.cn (D.W.); gyhhh_8010@163.com (Z.G.); mse_houpk@ujn.edu.cn (P.H.)

[2] College of Materials Science and Engineering, University of Jinan, Jinan 250022, Shandong, China; mse_yangp@ujn.edu.cn

* Correspondence: mse_chengx@ujn.edu.cn (X.C.); mse_huangsf@ujn.edu.cn (S.H.)

Received: 13 January 2020; Accepted: 28 January 2020; Published: 31 January 2020

Abstract: Surface application of photocatalyst in cement-based materials could endow it with photocatalytic properties, however, the weak adhesion between photocatalyst coatings and the substrates may result in poor durability in outdoor environments. In this study, TiO$_2$@SiO$_2$ core-shell nanocomposites with different coating thicknesses were synthesized by varying the experiment parameters. The results indicate that SiO$_2$ coatings accelerated the rhodamine B removal to a certain extent, owing to its high surface area; however, more SiO$_2$ coatings decreased its photocatalytic efficiencies. The cement matrix treated with TiO$_2$@SiO$_2$ core-shell nanocomposites showed good photocatalytic efficiency and durability after harsh weathering processing. A reaction mechanism was revealed by the reaction of TiO$_2$@SiO$_2$ nanocomposites with Ca(OH)$_2$.

Keywords: photocatalysis; core-shell structure; TiO$_2$/SiO$_2$ composite nanoparticles; cement-based materials

1. Introduction

The application of TiO$_2$ nanoparticles in cementitious materials has gained considerable interest and research, which could endow it with photocatalytic properties, such as self-cleaning [1], air-purifying [2], and antibacterial properties [3]. P25, a commercial TiO$_2$ nanoparticle, was widely employed in building materials, owing to its small size, excellent photocatalytic efficiency, being toxic-free, accessible, and affordable merits [4,5]. Poon [6] reported that P25 was used to modify the concrete surface of self-compacting mortars, and it was found to be effective in rhodamine B (RhB) degradation under UV and strong halogen light irradiation. Sánchez [7] mixed P25 into cement mortar using different dosages, and the mortars show outstanding degradation of NO$_x$ gases.

Traditionally, photocatalytic materials are introduced into cement-based materials via a direct mixing method. The obtained TiO$_2$-containing cement-based materials show accelerated cement hydration [8], lower porosity, improved mechanical properties [9], self-cleaning [10], air-purifying [11], and antibacterial properties [3]. However, the incorporation method faces a serious restriction in having limited photocatalytic efficiency and some conditions. Boule [12] studied TiO$_2$-containing cement-based materials prepared using different techniques, and the results show that the TiO$_2$-cement mixture has significantly less efficiency than TiO$_2$ slurries for the degradation of 3-nitrobenzenesulfonic acid and 4-nitrotoluenesulfonic acid. This is attributed to the reduction in the active surface. Meanwhile, the incorporation method is not applied in existing buildings and natural stone structures.

Enriching the surface of cement-based materials with a photocatalyst seems to be a viable option. However, the weak adhesion of the photocatalytic material to its substrate remains a big concern [13].

When the photocatalyst is exposed to harsh weather, the issue of adhesion gains more importance than it deserves. A binder of cement-based materials with a photocatalyst could be considered in order to improve durability. SiO_2 as an inorganic material is a valuable resource to prevent the release of the TiO_2 photocatalyst from the surface of cement-based materials to the environment, owing to its pozzolanic activity. Meanwhile, the adding of SiO_2 as a heterogeneous catalysis, such as $TiO_2@SiO_2$ and $SiO_2@TiO_2$, may improve the photocatalytic efficiency in comparison to pure TiO_2 [14–16]. Stephan [17] obtained silica-titania core-shell composites, and the results show that some core-shell particles have higher photocatalytic efficiency than pure nano-titania photocatalysts. Mendoza [18] studied a cement matrix treated with TiO_2 and TiO_2–SiO_2, which showed high RhB photodegradation conversions (above 80%). Sikora [19] reported that silica/titania ($mSiO_2/TiO_2$) core-shell nanocomposites improved the compressive strength, reduced the water absorption of cement mortars, and exhibited relatively good bactericidal properties.

In this study, $TiO_2@SiO_2$ core-shell composites with different coating thicknesses were designed and synthetized. SiO_2 coating accelerated RhB removal to a certain extent, and improved photocatalytic durability by test efficiencies before and after curing and an accelerated weathering process. A reaction mechanism was revealed by the reaction of $TiO_2@SiO_2$ nanocomposites with $Ca(OH)_2$.

2. Experimental Details

2.1. Materials

White Portland cement (WC, P.W 42.5R) was provided by the Aalborg cement company (Jinan, China), and the major chemical compositions of WC are shown in Table 1. Tetraethyl orthosilicate (TEOS), aqueous ammonia (25%), RhB, calcium hydroxide ($Ca(OH)_2$), and absolute ethanol of chemical grade were purchased from China National Pharmaceutical Group Corporation (Beijing, China).

Table 1. Major chemical compositions of withe Portland cement.

Constituent	CaO	SiO_2	SO_3	Al_2O_3	MgO	Fe_2O_3	TiO_2	K_2O	Na_2O
Mass (wt. %)	65.44	20.09	4.07	2.55	1.77	0.25	0.33	0.24	0.71

2.2. Synthesis of $TiO_2@SiO_2$ Nanocomposites

Commercial TiO_2 nanoparticles (P25) were used as the core structure, and the SiO_2 shell was obtained by the typial Stöber method. P25 (0.1 g) was dispersed into the solution of water (100 mL) and ethanol (80 mL) with an ultrasonic dispersion instrument for 30 min. Aqueous ammonia (1 mL) and TEOS (1 mL) were dropped above the mixture, respectively, under stirring for 8 h at 25 °C. $TiO_2@SiO_2$ nanocomposites were collected from the solution by a centrifugation method. All the experimental processes were the same, except for the experimental temperature, and nanocomposites obtained at 0 °C were prepared. Experimental parameters, such as temperature and the amount of P25, were adjusted to prepare different nanocomposites, and the detailed experimental data were listed in Table 2.

Table 2. Experimental parameters of $TiO_2@SiO_2$ nanocomposites.

Parameters TiO2@SiO2	P25 (g)	TEOS (mL)	Water (g)	Ethanol (mL)	NH3·H2O (mL)	Temperature (°C)	Main Thickness of the Shells (nm)	Surface Areas ($m^2 \cdot g^{-1}$)
Sample 1	0.05	0.3	100	80	1	25	3.92	96.63
Sample 2	0.10	0.3	100	80	1	25	2.25	46.63
Sample 3	0.15	0.3	100	80	1	25	1.95	46.49
Sample 4	0.05	0.3	100	80	1	0	6.13	110.42
Sample 5	0.10	0.3	100	80	1	0	4.17	99.75
Sample 6	0.15	0.3	100	80	1	0	3.70	91.42

2.3. Cement Paste Preparation, Curing and Surface Treatment

WC pastes (4 cm × 4 cm × 16 cm) were prepared, and the weight ratio of water to cement was 0.35. The paste specimens were put into a curing chamber at 95% relative humidity and 20 ± 2 °C for 28 days. After that, theses specimens were cut into slices with a size of 4 cm × 4 cm × 2 cm and then dried at 50 °C for 24 h. Pure P25 (0.025 g) and $TiO_2@SiO_2$ nanocomposites were dispersed into water (2 mL), respectively, and the dosages of nanocomposites were calculated to keep the same concentration of TiO_2, compared to pure P25 suspension. The obtained P25 and $TiO_2@SiO_2$ suspensions were sprayed on one surface (4 cm × 4 cm) of silices.

2.4. Characterization

Transmission electron microscope (TEM, FEI G2F20, Hillsboro, OR, USA) was employed to observe the morphologies of $TiO_2@SiO_2$ nanocomposites. Scanning electron microscope (SEM, ZEISS EVO LS15, Germany) was applied to obtain morphological images of the treated cement. N_2 adsorption-desorption isotherms were performed at −196 °C using a multi-function adsorption instrument (BET, Beijing Builder Company MFA-140, Beijing, China). The specific surface area was calculated by the Brunaur–Emmett–Teller method. Solid-state ^{29}Si MAS NMR spectra (Bruker AV600, Germany) were acquired on a Bruker AV600 spectrometer.

2.5. Photocatalytic Degradation of RhB

The test of RhB degradation was carried out to evaluate the self-cleaning performance of $TiO_2@SiO_2$ and coated WC pastes. In photocatalytic tests of $TiO_2@SiO_2$ powders, P25 (0.01 g) and nanocomposites were dispersed in an RhB solution (20 mL, 10 mg/L), and the TiO_2 dosages of nanocomposites kept the same concentration. The suspension was further stirred for 30 min in the dark and was then irradiated by a UV lamp (20 W). Part of the solution (2 mL) was taken every 15 min until 105 min, and was centrifuged to remove nanocomposites. The concentration of the obtained solution was tested using a UV/Vis spectrometer at a wavelength of 463 nm to investigate the degradation efficiency. To measure the RhB degradation of coated WC pastes, the surface coated with nanocomposites was sprayed with 2 mL of RhB solution (80 mg/L). RhB contaminated specimens were irradiated by the UV lamp, and color variations (ΔE) before and after irradiation were recorded using a portable sphere spectrophotometer (RM200, X-Rite). The efficiencies of RhB removal was calculated by the comparison of the color variations before ($\Delta E(0)$) and after several hours ($\Delta E(t)$) of UV light irradiation. The detailed calculation method was mentioned in References [20,21].

A lab-simulated weathering system was used to mimic rain water in an outdoor environment. Circular tap water with a flow rate of 130 mm/h was produced by pumping it from a water tank in order to simulate the rain condition, and the process lasted 5 days.

3. Results and Discussion

$TiO_2@SiO_2$ nanocomposites with different coating thicknesses were prepared to investigate its effect on photocatalytic efficiency and the bonding force with the cement-based matrix. In order to obtain $TiO_2@SiO_2$ nanocomposites with different coating thicknesses, experimental parameters, such as the temperature and dosage ratio of TEOS and P25, were adjusted to prepare different nanocomposites. Samples 1, 3, 4, and 6 were selected to investigate the effect of experimental parameters on the coating thicknesses of $TiO_2@SiO_2$ nanocomposites. The TEM images of these nanocomposites are shown in Figure 1. Compared to the TEM images of four nanocomposites, the SiO_2 shell of Samples 1 and 4 is obviously present. The results indicate that the dosage ratio of TEOS and P25 decides the molar ratio of SiO_2 and TiO_2, and the higher value promotes the formation of the SiO_2 shell. The enlarged insets of Figure 1a,c and show that the two nanocomposites both have core-shell structures, and the main shell thicknesses of Samples 1 and 4 are about 3.92 and 6.13 nm, respectively. During the preparation process of the coating, the rates of hydrolysis and condensation processes decreased at lower temperatures, and SiO_2 preferentially deposited on the surface of TiO_2 particles, benefiting the process of more coating.

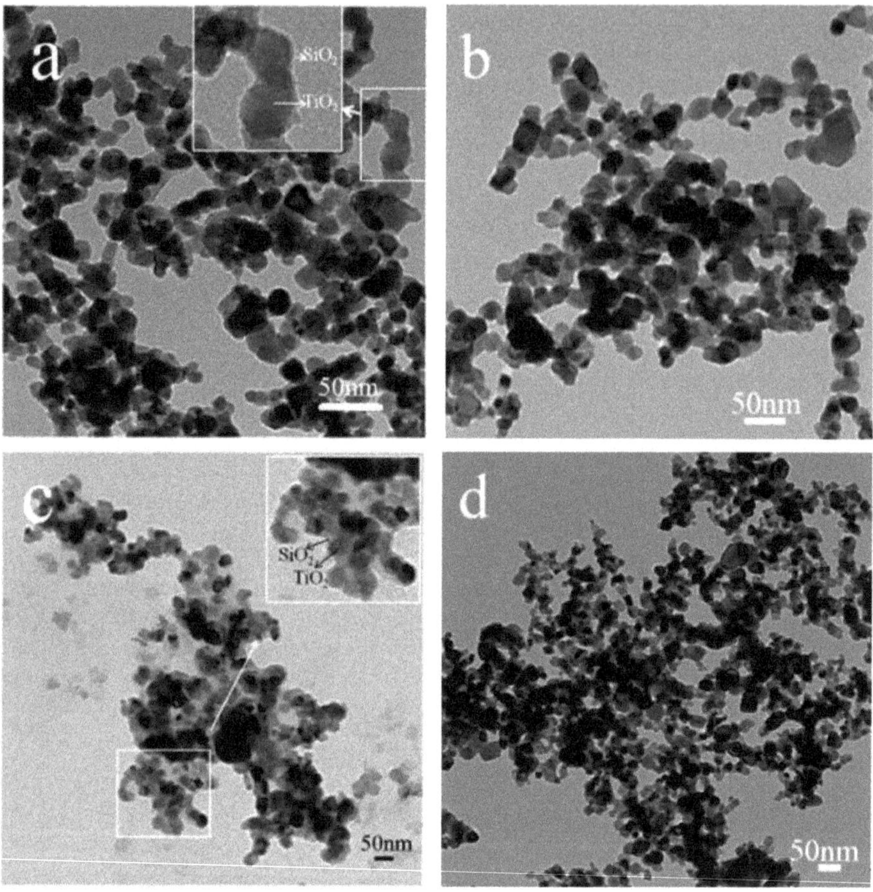

Figure 1. TEM images of Sample 1 (**a**), Sample 3 (**b**), Sample 4 (**c**), and Sample 6 (**d**).

The photocatalytic activities of $TiO_2@SiO_2$ nanocomposites with different shell thicknesses and P25 were evaluated by monitoring the degradation of RhB under UV-light irradiation, and the results are shown in Figure 2. After dark reaction for 30 min, the degradation rates of $TiO_2@SiO_2$ nanocomposites were higher than that of pure P25. Meanwhile, the values increase with the increase in the shell thickness, mainly owing to the absorption of the SiO_2 coating. Samples 2, 3, and 6 have higher degradation efficiencies than pure P25 when the irradiation time was 105 min, indicating that the SiO_2 coating accelerates RhB removal to a certain extent. This may be owing to the high surface area of SiO_2, which adsorbed more dyes to the benefit of more dye degradation [14]. Samples 1 and 4 have higher SiO_2 coating thicknesses, however, their photocatalytic efficiencies are the lowest. This is probably because much of the coating may hinder the transport of photons and decrease light absorption [18]. During the preparation process of $TiO_2@SiO_2$ nanocomposites, temperature may be an influencing factor for the thickness and density of the coating, which further affects the photocatlytic activity of nanocomposites. The nanocomposites prepared at 25 °C have higher degradation rates than nanocomposites synthetized at 0 °C, which may be due to the difference in the coating thickness and morphology.

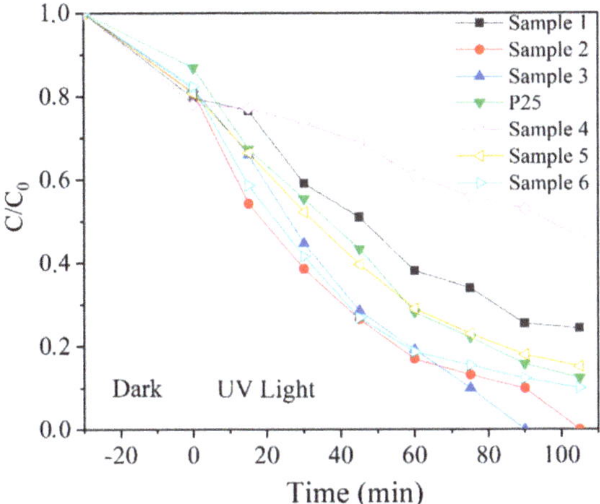

Figure 2. Photocatalytic degradation of RhB under UV light by P25 and TiO_2@SiO_2 nanocomposites.

Samples 1–3 and P25 were sprayed on the surface of WC pastes, and the specimens coated with Samples 1–6 and P25 were named WC 1–6 and WC P25, respectively. The surface microstructures are shown in Figure 3. The figures indicate that the surfaces of hardened cement pastes treated with Samples 1–3 are denser than those of untreated cement pastes and the paste treated with P25. The results may be due to the gelatinization of SiO_2 or the pozzolanic reaction of SiO_2 and cement matrix. The adding of SiO_2 shell improves the surface quality of cement-based materials, and it may protect against the release of TiO_2 in a harsh weathering environment. In order to investigate the photocatalytic properties of WC specimens, the RhB degradation rates of these specimens were first measured after the TiO_2@SiO_2 nanocomposites were sprayed on the surface of the WC pastes. After that, these specimens were put into a curing chamber for two weeks. TiO_2@SiO_2 nanocomposites may react with the WC matrix due to their pozzolanic activity, and the degradation rates were measured again to investigate the effect of reaction productions on the photocatalytic efficiency. If the pozzolanic reaction occurs, the production could decrease the release of TiO_2 after exposure to a harsh weathering process. Therefore, after the weather process, degradation rates of specimens were tested. The results are shown in Figure 4, and Table 3 shows the detailed data from the figure. RhB degradation rates of WC 1 and WC 4 are the highest, and the tendency is different from the degradation rates of TiO_2@SiO_2 nanocomposites. This probably because adsorption action is more effective on the RhB degradation after nanocomposites are applied on the surface of WC pastes. After curing, their degradation rates significantly increase. C-S-H gel with relatively larger BET surface areas may be formed due to the pozzolanic reaction, which may adsorb more RhB dye. The results indicate that the reaction production has no negative effect on photocatalytic efficiency. Instead, it increases BET surface areas to promote RhB degradation. The production may decrease the release of TiO_2 and improve the durability of photocatalytic activity. After the weather process, the RhB removals of all specimens were decreased, and the reduction of WC P25 was obvious. WC 1 and WC 4 show the minimum photocatalytic efficiencies, which was due to more SiO_2 coating of Samples 1 and 4. Meanwhile, WC 4 has the minimum value due to its larger thickness of the SiO_2 coating mentioned above. The results show that adding SiO_2 could improve the adhesion between the coating and substrates, and the durability of the photocatalytic property, in particular for a harsh weathering process. The RhB degradations of WC 2 and WC 3 are similar, mainly owing to their similar BET surface areas, as mentioned above.

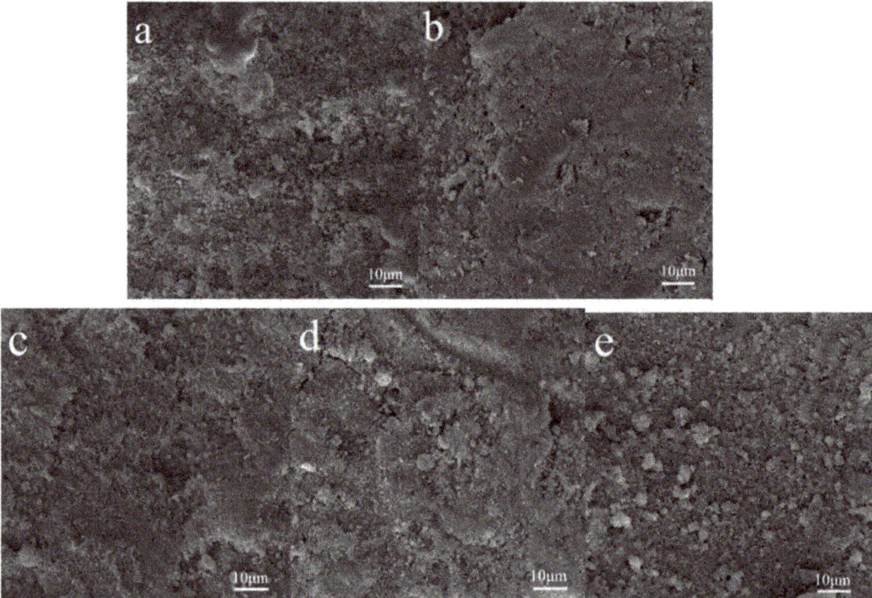

Figure 3. SEM images of harden cement pastes (**a**) and samples treated with Samples 1–3 (**b–d**) and P25 (**e**).

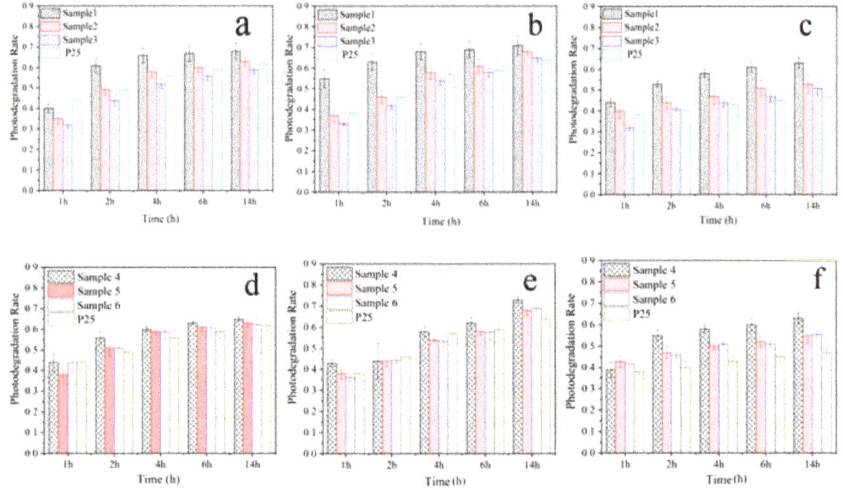

Figure 4. RhB degradation rates of WC 1-6 and WC P25 before (**a,d**) and after curing (**b,e**) and accelerated weather process (**c,f**).

Table 3. RhB degradation rates and reduction values of WC 1-6 and WC P25 after irradiation.

Sample	Sample 1	Sample 2	Sample 3	Sample 4	Sample 5	Sample 6	P25
Direct measurement	68.1%	63.1%	59.4%	65.0%	63.1%	62.5%	62.4%
After curing	71.5%	68.1%	65.0%	73.4%	68.0%	69.1%	64.0%
After weather process	62.9%	53.4%	51.0%	63.3%	55.0%	55.6%	47.4%
Reduction	7.6%	15.4%	14.1%	2.6%	12.8%	11.0%	24.0%

To more specifically study the pozzolanic reaction of $TiO_2@SiO_2$ nanocomposites with the cement matrix, experiments of nanocomposites with $Ca(OH)_2$ (a hydration product of cement) were carried out to investigate the formation of the reaction production. Sample 2 was selected and was mixed with $Ca(OH)_2$ saturated solution by stirring for seven days. TEM image of the reaction product of Sample 2 and $Ca(OH)_2$ is shown in Figure 5. A foil-like structure was formed after the reaction, which may be a C-S-H gel. The particles are deposited on the surface of a foil-like structure. The SiO_2 shell reacted with $Ca(OH)_2$, and the product core was first deposited on the particles. The product core further grew to a foil-like structure. This may explain why the adding of SiO_2 decreased the release of TiO_2 and improved the durability of the photocatalytic activity.

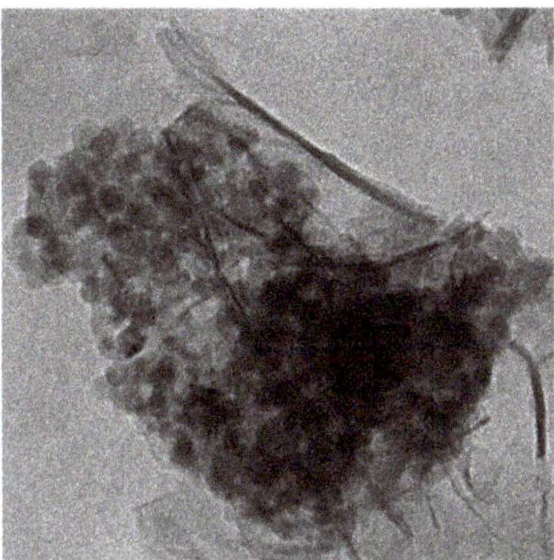

Figure 5. TEM image of reaction product of Sample 2 and $Ca(OH)_2$.

The objective of ^{29}Si NMR spectrum is to assess the composition of the foil-like structure and Q_n structure of the Si tetrahedron. Figure 6 shows the spectra of Sample 2 and its reaction production with $Ca(OH)_2$. The spectrum of Sample 2 shows two signals centered at 102.5 and 113.1 ppm, which were associated with the Q_3 and Q_4 units in the SiO_2 structure, according to previous reports [22,23]. After the reaction with Sample 2 and $Ca(OH)_2$, the two main peaks shifted to 81.7 and 88.0 ppm, which were assigned to Q^1 and Q^2 silicate connections. The results indicate a decrease in the polymerization degree of silicon and the formation of a C-S-H gel. After the reaction, the SiO_2 coating transformed to a C-S-H gel, which could decrease the release of TiO_2 after exposure to a harsh weathering process.

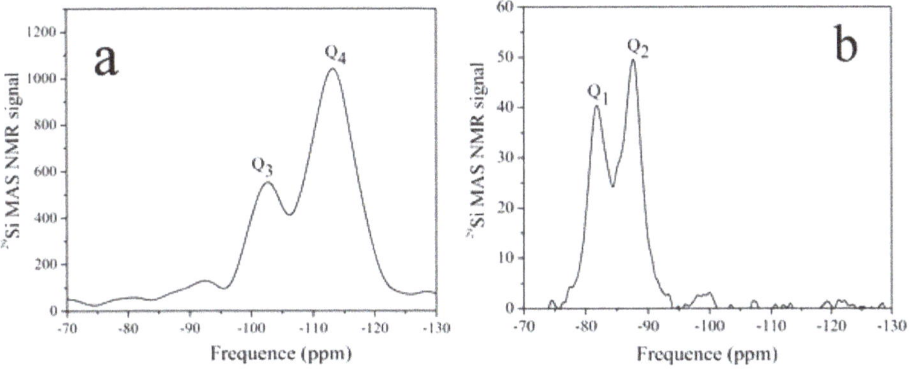

Figure 6. NMR spectra of Sample 2 (**a**) and its reaction production (**b**) with $Ca(OH)_2$.

4. Conclusions

$TiO_2@SiO_2$ nanocomposites were prepared using different experiment parameters, and their RhB degradations were measured. The results show that SiO_2 coating accelerates the RhB removal to a certain extent, owing to the high surface area of SiO_2, which adsorbed more dyes to the benefit of more dye degradation. More SiO_2 coating may hinder the transport of photons and decrease light absorption, which further decreases photocatalytic efficiencies. After these nanocomposites were applied on the surface of the WC pastes, their RhB degradation rates are higher than WC P25, which is probably because adsorption action is more effective on the RhB degradation. After curing, their degradation rates significantly increase. C-S-H with relatively large BET surface areas may be formed due to the pozzolanic reaction, which may adsorb more RhB dye. After the weather process, the RhB removals of all specimens decreased, and the reduction of WC P25 was obvious. Meanwhile, WC 4 has the minimum value. Adding SiO_2 could improve the adhesion between coating and substrates, and the durability of the photocatalytic property, in particular for a harsh weathering process. The experimental data of nanocomposites with $Ca(OH)_2$ further prove the formation of C-S-H gels and reveal the reaction of the mechanism of nanocomposites and the cement matrix.

Author Contributions: X.C. and D.W. conceived and designed the study; D.W. and Z.G. performed the experiments; D.W. wrote the paper; P.Y., S.H., and X.C. reviewed and edited the manuscript. Resourse: P.H. All authors have read and agreed to the published version of the manuscript.

Funding: This research was funded by the Program for Taishan Scholars Program, Case-by-Case Project for Top Outstanding Talents of Jinan, Distinguished Taishan Scholars in Climbing Plan, National Natural Science Foundation of China (Grant No. 51632003 and 51902129), the National Key Research and Development Program of China, (Grant No. 2016YFB0303505), and the 111 Project of International Corporation on Advanced Cement-based Materials (No. D17001).

Acknowledgments: This work was supported by the Program for Taishan Scholars Program, Case-by-Case Project for Top Outstanding Talents of Jinan, Distinguished Taishan Scholars in Climbing Plan, National Natural Science Foundation of China (Grant No. 51632003 and 51902129), the National Key Research and Development Program of China, (Grant No. 2016YFB0303505), and the 111 Project of International Corporation on Advanced Cement-based Materials (No. D17001).

Conflicts of Interest: The authors declare no conflict of interest.

References

1. Guo, M.Z.; Maury-Ramirez, A.; Poon, C.S. Self-cleaning ability of titanium dioxide clear paint coated architectural mortar and its potential in field application. *J. Clean Prod.* **2016**, *112*, 3583–3588. [CrossRef]
2. Chen, M.; Chu, J.W. NO_x photocatalytic degradation on active concrete road surface-from experiment to real-scale application. *J. Clean. Prod.* **2011**, *19*, 1266–1272. [CrossRef]
3. Guo, M.Z.; Maury-Ramirez, A.; Poon, C.S. Versatile photocatalytic functions of self-compacting architectural glass mortars and their inter-relationship. *Mater. Des.* **2015**, *88*, 1260–1268. [CrossRef]
4. Yousefi, A.; Allahverdi, A.; Hejazi, P. Effective dispersion of nano-TiO_2 powder for enhancement of photocatalytic properties in cement mixes. *Constr. Build Mater.* **2013**, *41*, 224–230. [CrossRef]
5. Cárdenas, C.; Tobón, J.I.; García, C.; Vila, J. Functionalized building materials: Photocatalytic abatement of NOx by cement pastes blened with TiO2 nanoparticles. *Constr. Build Mater.* **2012**, *36*, 820–825.
6. Chen, J.; Kou, S.C.; Poon, C.S. Photocatalytic cement-based materials: Comparison of nitrogen oxides and toluene removal potentials and evaluation of self-cleaning performance. *Build Environ.* **2011**, *46*, 1827–1833. [CrossRef]
7. Sugrañez, R.; Álvarez, J.I.; Cruz-Yusta, M.; Mármol, I.; Morales, J.; Vila, J.; Sánchez, L. Enhanced photocatalytic degradation of NOx gases by regulating the microstructure of mortar cement modified with titanium dioxide. *Build Environ.* **2013**, *69*, 55–63. [CrossRef]
8. Sun, J.F.; Xu, K.; Shi, C.Q.; Ma, J.; Li, W.F.; Shen, X.D. Influence of core/shell $TiO_2@SiO_2$ nanoparticles on cement hydration. *Constr. Build Mater.* **2017**, *156*, 114–122. [CrossRef]
9. Tiainen, H.; Wiedmer, D.; Haugen, H.J. Processing of highly porous TiO_2 bone scaffolds with improved compressive strength. *J. Eur. Ceram Soc.* **2013**, *33*, 15–24. [CrossRef]

10. Ruot, B.; Plassais, A.; Olive, F.; Guillot, L.; Bonafous, L. TiO_2-containing cement pastes and mortars: Measurements of the photocatalytic efficiency using a rhodamine B-based colourimetric test. *Sol. Energy* **2009**, *83*, 1794–1801. [CrossRef]

11. Folli, A.; Pade, C.; Hansen, T.B.; Marco, T.D.; Macphee, D.E. TiO_2 photocatalysis in cementitious systems: Insights into self-cleaning and depollution chemistry. *Cem. Concr. Res.* **2012**, *42*, 539–548. [CrossRef]

12. Rachel, A.; Subrahmanyam, M.; Boule, P. Comparison of photocatalytic efficiencies of TiO_2 in suspended and immobilised form for the photocatalytic degradation of nitrobenzenesulfonic acids. *Appl. Cat. B Environ.* **2002**, *37*, 301–308. [CrossRef]

13. Martinez, T.; Bertron, A.; Escadeillas, G.; Ringot, E.; Simon, V. BTEX abatement by photocatalytic TiO_2-bearing coatings applied to cement mortars. *Build Environ.* **2014**, *71*, 186–192. [CrossRef]

14. Kamaruddin, S.; Stephan, D. Quartz-titania composites for the photocatalytical modification of construction materials. *Cem. Concr. Comp.* **2013**, *36*, 109–115. [CrossRef]

15. Kamaruddin, S.; Stephan, D. Sol-gel Mediated Coating and Characterization of Photocatalytic Sand and Fumed Silica for Environmental Remediation. *Water Air Soil Pollut.* **2014**, *225*, 1948. [CrossRef]

16. Hendrix, Y.; Lazaro, A.; Yu, Q.; Brouwers, J. Titania-Silica Composites: A Review on the Photocatalytic Activity and Synthesis Methods. *World J. Nano Sci. Eng.* **2015**, *5*, 161–177. [CrossRef]

17. Kamaruddin, S.; Stephan, D. The preparation of silica-titania core-shell particles and their impact as an alternative material to pure nano-titania photocatalysts. *Catal. Today* **2011**, *161*, 53–55. [CrossRef]

18. Mendoza, C.; Vallea, A.; Castellote, M.; Bahamondea, A.; Faraldosa, M. TiO_2 and TiO_2-SiO_2 coated cement: Comparison of mechanic and photocatalytic properties. *Appl. Catal. B Environ.* **2015**, *178*, 155–164. [CrossRef]

19. Sikora, P.; Cendrowski, K.; Markowska-Szczupak, A.; Horszczaruk, E.; Mijowska, E. The effects of silica/titania nanocomposite on the mechanical and bactericidal properties of cement mortars. *Constr. Build Mater.* **2017**, *150*, 738–746. [CrossRef]

20. Wang, D.; Hou, P.K.; Yang, P.; Cheng, X. $BiOBr@SiO_2$ flower-like nanospheres chemically-bonded on cement-based materials for photocatalysis. *Appl. Surf. Sci.* **2018**, *430*, 539–548. [CrossRef]

21. Wang, D.; Hou, P.K.; Zhang, L.N.; Yang, P.; Cheng, X. Photocatalytic and hydrophobic activity of cement-based materials from benzyl-terminated-TiO_2 spheres with core-shell structures. *Constr. Build Mater.* **2017**, *148*, 176–183. [CrossRef]

22. Vinogradova, E.; Estrada, M.; Moreno, A. Colloidal aggregation phenomena: Spatial structuring of TEOS-derived silica aerogels. *J. Colloid Interface Sci.* **2006**, *298*, 209–212. [CrossRef] [PubMed]

23. Humbert, B. Estimation of hydroxyl density at the surface of pyrogenic silicas by complementary NMR and Raman experiments. *J. Non-Cryst. Solids* **1995**, *191*, 29–37. [CrossRef]

Article

Experimental Investigations on the Performances of Composite Building Materials Based on Industrial Crops and Volcanic Rocks

Raluca Iștoan *, Daniela Roxana Tămaș-Gavrea and Daniela Lucia Manea

Department of Civil Engineering and Management, Faculty of Construction,
Technical University of Cluj-Napoca, 28 Memorandumului Street, 400114 Cluj-Napoca, Romania;
roxana.tibrea@ccm.utcluj.ro (D.R.T.-G.); daniela.manea@ccm.utcluj.ro (D.L.M.)
* Correspondence: ralucafernea@ccm.utcluj.ro

Received: 16 January 2020; Accepted: 8 February 2020; Published: 11 February 2020

Abstract: Interdisciplinary and sustainability represent the main characteristics of this paper due to the fact that this research is offering a connection between two main areas—agronomy and construction, by using hemp shiv for the design of new building materials, which can increase the sustainability level of the building industry. For this reason, the main scope of this study is based on the investigation of a new category of composite building materials—lightweight mortars based on hemp shiv, volcanic rocks and white cement—which contribute to a positive environmental impact and help to increase indoor comfort. A complex report was carried out on two segments. The first one is focused upon the characteristics of the raw materials from the composition of the new materials, while the second segment presents a detailed analysis of these composites including morphological and chemical investigation, pyrolytic and fire behavior, compression and flexural strengths, and acoustic and thermal characteristics. The proposed recipes have as a variable volcanic rocks, while the hemp and the binder maintain their volumes and properties. The results were analyzed according to the influence of volcanic rocks on the new composites.

Keywords: sustainability; hemp; sound absorption; thermal conductivity; fire characteristics; mechanical properties; microscopy; thermal analysis; X-ray diffraction

---

1. Introduction

Sustainability and building materials are two different parts that together can contribute to a new perspective on the building industry. Constructions today play an important and critical role in developing a sustainable society [1]. For this reason, sustainable development represents a new challenge for each sector to become environmentally-friendly, by protecting the planet's limited resources and preventing additional pollution. During the last years of activity, the construction sector has supported the tendency to build in a more responsible way, so far as environmental and human life protection are concerned. New keywords define buildings such as passive, green, eco-friendly, sustainable and so forth. Different natural and traditional resources such as straw bale [2], hemp [3], flax [4], cob [5], sheep wool [6] and so forth are reconsidered as building materials. Moreover, new composite materials based on waste, such as glass [7,8], rubber [9] and so forth, have been studied and investigated. The use of the hemp shiv as a raw building material is not new; it was rediscovered after 1990 [10]. Its main characteristics concern improving sound and thermal performance, and having a positive impact on the environment, air-permeability, preventing condense formation and waterproofing [11]. New approaches with hemp shiv—which is regarded as a waste product obtained after the extraction of the fibers for the textile industry—have been taken into consideration in the construction sector. There is a significant number of studies on hemp shiv with lime, which is a composite material named hempcrete [12–14], but there are only few studies that concern hemp shiv with cement [15–19].

Due to the fact that hemp shiv presents poor fire resistance, a new approach can be made with volcanic rocks, which are natural products with higher fire resistance [20]. These materials are usually found in the composition of mortars used in the rehabilitation process of the walls, because of their increased mechanical resistances [21]. They also present sound and thermal properties, as well as a low density, which will help the workability of the mortar. Durable and with higher strength properties for structural application [20], volcanic rocks are low cost materials [22] that also present fire retarding properties for steel structures [23].

Therefore, the purpose of this paper is to analyze the impact of the volcanic rocks when these are incorporated in hemp–cement composite materials used as interior mortars. These building materials are developed to be suitable not only for eco-friendly construction but also in regular buildings made from concrete, steel, and so forth, where they can increase indoor comfort.

2. Materials and Methods

2.1. Raw Materials

The composition of the new materials is designed to increase indoor comfort by developing new types of lightweight mortars, obtained as prefabricated plates. In order to understand the characteristics of the new composites, hemp shiv, natural and treated; volcanic rocks, perlite and vermiculite; and binders, white cement and hydrated lime, were analyzed.

2.1.1. Hemp Shiv

The hemp shiv used for the new composites was received from HempFlax Romania (Pianu De Jos, Alba, Romania). The chemical structure of the hemp shiv is represented by cellulose 44%, hemicellulose 18%–27%, lignin 22%–28%, extractives (fatty acids, waxes, sterols, triglycerides, steryl esters, glycosides, fatty alcohols, terpenes, phenolics, simple sugars, alkaloids, pectins, gums and essential oils) 1%–6% and ash 1%–2% according to Reference [24]. The percentages may differ depending on the hemp variety used. Hemp shiv presents a natural structure, shown in Figure 1a, similar to wood fibers, with dimensions from 1 to 8 mm, as follows: 4–8 mm (48.75%), 2–4 mm (22.5%), 1–2 mm (25%) and 0.25–1 mm (4.75%). The sorting procedure was made using the Controls sort system with 6 sieves of 0.25; 0.5; 1; 2; 4 and 8 mm. The relative humidity of the hemp shiv was approximately 7% and the density was 100 kg/m^3. The density was found to be similar to that found in the investigation of Reference [13]. The microscopic image of the hemp shiv presented in Figure 1b shows the cellular porous microstructure with two different sizes of longitudinal channels. According to References [25,26], the small ones range from 20–50 μm and are called xylem rays, whereas the bigger ones are on average around 50–100 μm and are named vessels. The surface walls of the vessel show perforations that permit a connection with the xylem ray, making it possible for the moisture to pass from one cell to the other. Regularly, the vessels are individually arranged and the pore frequency is about 20.8 vessels/mm^2 [26]. The thermal analyses of the natural hemp shiv, presented in Figure 1c, showed a total degradation of 81.20%. The hemp shiv lost 3.6% on the first temperature interval from 20–200 °C and 77.6% on the second one, from 200 and 1000 °C. Compared with References [27,28], the second temperature interval is defined by the thermal combustion of hemicellulose and cellulose. The X-ray diffraction (XRD) of the hemp shiv was presented in a previous study [19], where the higher peak was recorded at 2 θ = 22.66°, similar to the studies of References [25,29]. The peak represents the crystallographic plane of the cellulose. Because of the similarity with the other two studies mentioned before, the hemp shiv presents crystalline and amorphous phases. The cellulose represents the crystalline part of the hemp shiv. The hemicellulose and lignin make the amorphous part, due to the hydroxyl group [25,29].

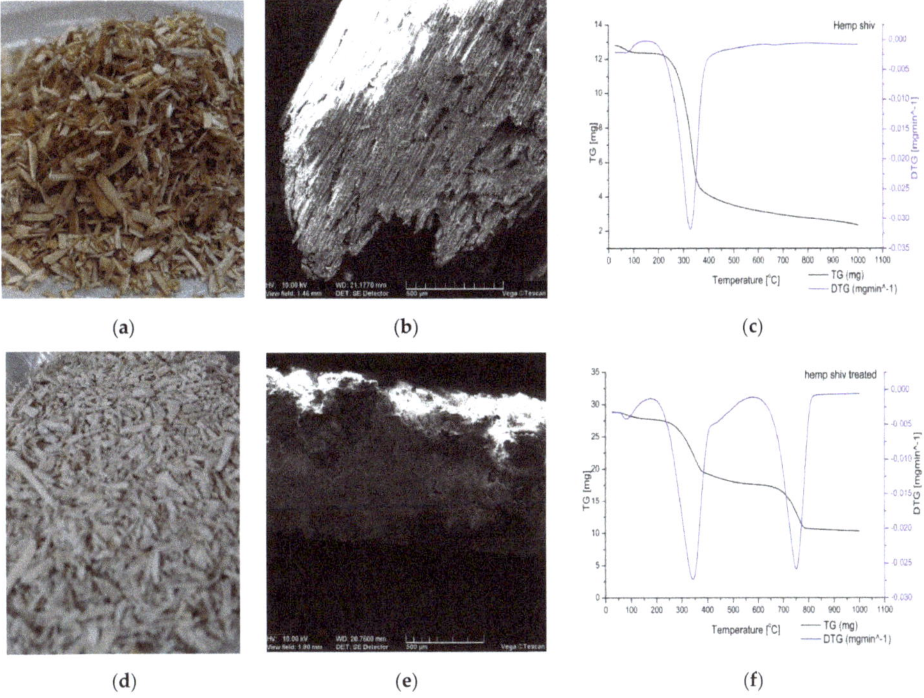

Figure 1. Industrial crop (**a**) natural hemp shiv; (**b**) natural hemp microscopy; (**c**) natural hemp thermal analysis; (**d**) treated hemp shiv; (**e**) treated hemp microscopy; (**f**) treated hemp thermal analysis.

The hemp shiv used in the recipes was treated with a lime solution, intended to improve the antiseptic and the fire characteristics of biomass. The treatment procedure consisted of soaking the hemp shiv under agitation for 5 min in a lime milk solution, until all the hemp shiv was covered by the solution. Successively, the hemp was dried under normal laboratory conditions at 24 °C and a humidity of 40%. Applying the lime solution to the hemp did not modify the density of the fibers, as highlighted in Reference [30]. The density kept the same value of 100 kg/m^3. Instead, the relative humidity changed, by decreasing to 4%. Figure 1d shows the visual aspect of the hemp shiv covered with lime solution, meanwhile Figure 1e captures the microscopic surface of hemp shiv at a scale of 500 μm. The image presents similarities with the studies in References [31,32]. The thermal analyses of the treated hemp shown in Figure 1f showed an improvement of the pyrolytic behavior of this plant, the degradation percentage being 64.15%. By treating the hemp shiv with lime solution, the mass loss was smaller by 17% at higher temperatures, in comparison with the natural ones. The diagram of the treated hemp shiv showed three temperature intervals. The first one, from 20 to 200 °C, presented 4.24% degradation of the shiv which is attributed to the water released [33]. The deterioration from the second and the third temperature interval, from 200 to 580 °C (34.74%) and 580 to 100 °C (25.37%) is characteristic of the lime degradation process [12]. The X-ray diffraction diagram of the treated hemp shiv indicated the same shape of graph as with the natural hemp shiv from Reference [19]. The difference consisted in the identification of new crystalline compounds—Ca(OH)$_2$ calcium hydroxide and CaCO$_3$ calcium carbonate. The central peak remains in the same position but varies in intensity and secondary peaks appear where the main compound is made of the carbonyl group. A remark regarding these changes is made in Reference [30], which assumes that by applying the lime treatment a part of the amorphous components of hemp, such as hemicelluloses, pectins and waxes, are removed.

2.1.2. Volcanic Rocks

The volcanic rocks were procured by Procema Perlit (Jilava, Ilfov, Romania). The rocks were selected based on composition, to reinforce the fire behavior of the composite materials, but also to grow the porosity of the elements. They are natural products, with significant thermal properties due to the porous structure. The volcanic rocks, perlite and vermiculite were chosen for the new compositions. The two igneous rocks were obtained after the cooling of volcanic eruptions. The main constituents of the perlite rock are SiO_2, Al_2O_3, K_2O, Na_2O; when hydrated it can also contain TiO_2, CaO, MgO, and Fe_2O_3 in small quantities [34]. When it comes to vermiculite, the main components are SiO_2, Al_2O_3, Fe_2O_3, CaO, MgO, K_2O but also some small quantities of Li, Cr, Ti, and Ni [35].

Perlite has a granular aspect, as shown in Figure 2a, with white color and it is usually used in the constructions sector as a lightweight aggregate for mortars, screeds and thermal insulation concrete. According to the manufacturer's datasheet, perlite has a density around 100–140 kg/m^3 (SR EN 1097-3/2002), for a grain size between 0–2.5 mm (SR EN 933/2002) and a thermal conductivity of about 0.055 W/mK (SR EN 12667/2002). The natural humidity is about 1% (SR EN 1097-6/2002), the crushing resistance is 0.10 N/mm^2 (SR EN 13055/2003) and the fire reaction is class A1. The microscopic image presented in Figure 2b illustrates the exterior porous surface of a perlite grain. According to Reference [36], the unexpanded perlite can increase volume by 10 to 20 times more, when heated over 700 °C, becoming an expanded perlite. The water molecules from the initial stage evaporate during the heating process, cracks appear as a consequence of the steam leak, therefore the resulting cavities will define the perlite structure [34,36,37]. The X-ray diffraction graph of perlite, from Reference [19], resembles the curve shape of powder perlite studied in Reference [38], indicating an amorphous structure already identified in previous studies [39,40]. The crystalline phase of perlite is provided by the compound of SiO_2 and the amorphous structure is given by the aluminosilicate compounds [39].

Figure 2. Volcanic rocks (**a**) perlite natural structure; (**b**) perlite microscopy; (**c**) vermiculite natural structure; (**d**) vermiculite microscopy.

The expanded vermiculite is known as a granular product, of brown color, shown in Figure 2c, which has a mica aspect, and is rich in iron ions, magnesium and silicates. The physical characteristics of vermiculite presented in the manufacturer's datasheet, indicate a grain size from 0–0.5 mm (max. 5%) and 0.5–3 mm (95%), a density of around 110–130 kg/m^3, whereas the absorption is around 60%–70% of the volume. It is usually used in horticultural substrates. Therefore, by including this product in the recipe, it was intended to analyze its behavior in composite materials. The microscopic image of the vermiculite grain captured in Figure 2d showed a lamellar structure formed during the speedy heating process [20,37,41–43]. The X-ray diffraction of vermiculite [19] shows a high peak with an intensity of 800 and a diffraction angle of 2 θ = 27.4°. The crystallographic determination of expanded vermiculite showed that this volcanic rock presents an incomplete crystallinity, having a considerable number of amorphous compounds [20,42]; therefore, it resembles the perlite. The thermal analysis of the volcanic rocks cannot be shown due to the fact that some errors appeared in the investigation process. Some studies from the scientific literature showed that the degradation process of the perlite is defined by three temperature intervals with a maximum percentage of degradation around 2% [44,45]. Reported to the vermiculite, the degradation process is around 9% [20]. These characteristics of the volcanic rocks, that is, resistance to higher temperatures, will help decrease the degradation process of the composite hemp based materials.

2.1.3. Binders

The hydrated lime used for the treatment of the hemp shiv was CL80-S according to EN 459-1, produced by Carmeuse (Brasov, Romania). The chemical composition of the lime powder, according to the manufacturer's datasheet, is CaO + MgO (min.80%), CO_2 (max.15%), MgO (max.5%) and SO_3 (max. 2%). Other laboratory tests concerned density (about 520 kg/m^3), porosity (74%) and compactness (26%). Lime powder is presented in Figure 3a, and the microscopy for this power at the scale of 500 μm can be observed in Figure 3b. The thermal analysis is presented in Figure 3c. The maximum percentage of degradation recorded by lime powder was 35.11%. The decomposition analysis was defined by two intervals of temperature. The first one, between 20–540 °C, where the sample recorded a decrease of mass by 11% while on the second temperature interval of 540–1000 °C was 24%. For the X-ray diffraction [19] the graph form showed only the presence of crystalline compounds—calcium hydroxide and calcium carbonate—which are characteristic for lime [38,46].

White cement was chosen with the main purpose of increasing the mechanical and fire characteristics of the new composite materials, but at the same time to offer a proper aesthetic side. The Portland white cement CEM I 52.5 produced by Devnya Cement was used, according to the BSS EN 197-1:2011. The cement powder is characterized by a white color, with a more intense brightness than the hydrated lime, seen when comparing Figure 3d with Figure 3a. The microscopic representation for the white cement powder was captured at the scale of 500 μm, shown in Figure 3e. The decomposition process of the cement powder was determined with the thermal analysis investigation. The maximum percentage of lost mass by the samples was 4.35%. The degradation process was defined by three temperature intervals: 0–240 °C (1.08%), 240–500 °C (0.70%) and 500–1000 °C (2.56%). The X- ray diffraction [19] indicated high contents of CaO and SiO_2, which define the main crystalline compounds of the cement—tricalcium silicate and dicalcium silicate—while the Al is found in small quantities, which determine the presence of the mineralogical compound of tricalcium aluminate, confirmed also by References [47–50]. Also, the X-ray diffraction of the white cement captures the presence of TiO_2, which is responsible for the white pigment of the Portland cement [47].

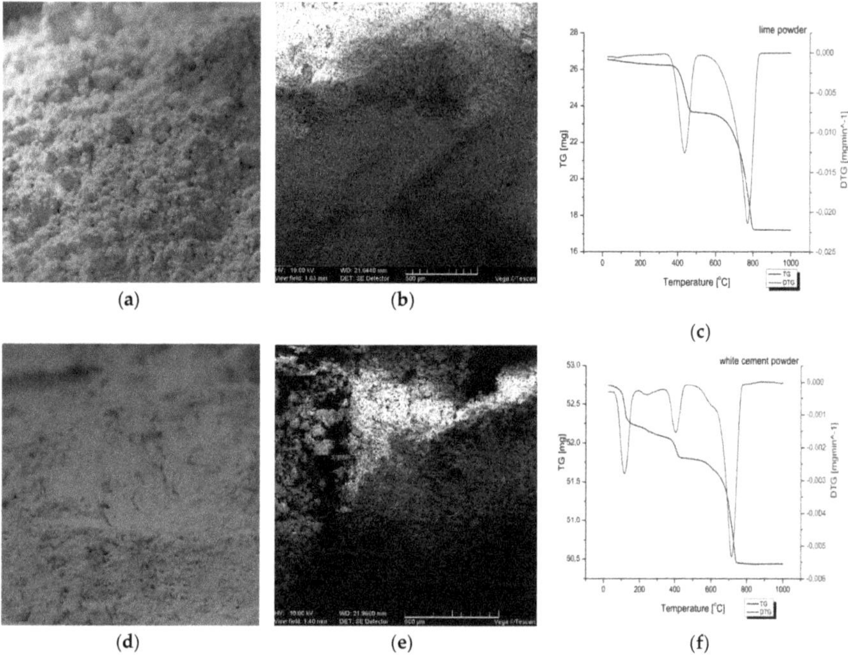

Figure 3. Binders (**a**) lime powder; (**b**) lime microscopy; (**c**) lime thermal analysis; (**d**) white cement powder; (**e**) white cement microscopy; (**f**) white cement thermal analysis.

2.2. Preparation of New Composite Materials

The mixing proportions of the new mortars were set up based on a previous study [18]. The composition of the analyzed samples started from the standard cement mortar formula. The sand grains were replaced with hemp shiv, obtaining three new compositions with a different hemp shiv volumes. The conclusion of the previous paper showed that the mix with a higher quantity of hemp has better acoustical and thermal characteristics. Therefore, to improve the obtained characteristics of this composition, a new set of recipes was analyzed. The new element introduced in the mortar mix was the volcanic rock, aiming to increase the fire properties of the materials, but also to obtain a lighter structure. Two types of volcanic rocks were proposed; perlite and vermiculite. Based on these new elements, three different compositions were analyzed, according to Table 1.

Table 1. Mixing proportion.

Identification Name in the Text Composition Name	M1 $C + C_3 + P_2$	M2 $C + C_3 + V_2$	M3 $C + C_3 + PV_2$
Ratio by Volumes			
White cement (C)	1	1	1
Hemp shiv (C)	3	3	3
Perlite (P)	2	-	1
Vermiculite(V)	-	2	1
Water	1	1	1

The dosages of the three mixtures are expressed in volumes. The ratio between binder and hemp was 1:3, the cement to volcanic rocks ratio was 1:2 and the binder to water ratio was 1:1. The mixing sequence followed the traditional part of mortar recipe, by adding the volcanic rocks at the end of the process. Mixing was performed with a mechanical mixer. The composition was placed into the molds in a single layer and manually compacted. For each determination, different molds were used as required by the test procedure. The samples were kept in the molds for 24 h in laboratory conditions with a temperature of 24 °C and a humidity of 40%. The three compositions were tested after 28 days.

2.3. Visual Analysis

The visual aspect of the compositions is shown in Figure 4. The mixtures present as homogeneous, with a uniform distribution of hemp fibers. Being designed for building interior mortars, white cement was chosen as a binder in order to comply with the aesthetic requirements. The compositions have different shades, depending on the volcanic rock used; therefore, the perlite compositions have a lighter shade than those with vermiculite, which, due to the mica appearance of the rock, are defined by a brown color.

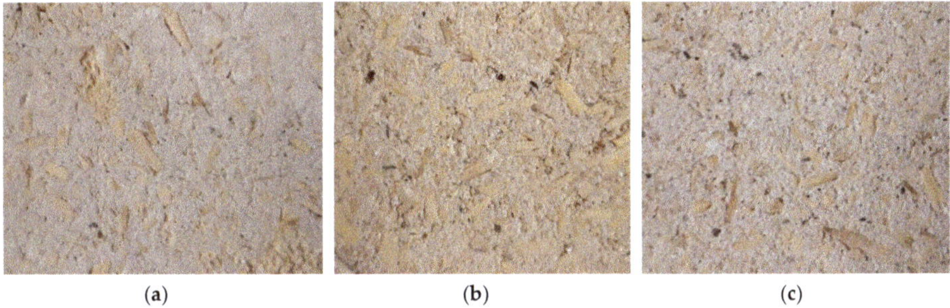

(a)   (b)   (c)

Figure 4. Visual composition (**a**) M1 (C + C_3 + P_2); (**b**) M2 (C + C_3 + V_2); (**c**) M3 (C + C_3 + PV_2).

2.4. Chemical Characterization

The modal composition of the samples was determined by using a non-destructive method called X-ray diffraction. The technique was helpful in the identification and quantification of crystalline compounds. The samples were analyzed in the form of powder. The method consisted of recording the reflected radiation after the propagation of X-rays on the surface of the sample with a fixed wavelength and intensity. Depending on the arrangement of the atoms in the structure, the materials were classified into crystalline and noncrystalline compositions. The crystallinity of the materials is given if the atoms are disposed of in a repeated way over large atomic distances while the amorphous compounds are defined by a random display of the atoms [51]. The crystallinity of the samples was measured using a Shimadzu 6000 XRD diffractometer. The diffraction angle used was between $2\,\theta = 10$–$70°$ and the applied wavelength was equal to $\lambda = 1.54182$ Å.

2.5. Morphological Analysis

The microstructural morphology of the samples was evaluated with a Scanning Electron Microscope, model VEGA, and with the software Tescan. The samples were broken into small pieces and, before the scanning process, they were sputter-coated with a layer of gold, with a thickness of 10nm, using the turbo molecular pumped coating system Q150T ES.

2.6. Thermogravimetry Analysis

The thermal analysis of the composite materials was proposed due to the fact that the ratio between hemp and binder was 3:1. The large volume of hemp from the compositions required more attention on the pyrolitic behavior of the existing biomass. Thermogravimetry analysis (TGA) is a method which presents the modifications in materials, both at physical or chemical levels, due to the change of temperature [52]. The analysis was carried out using a TGA/SDTA 851e_METTLER TOLLEDO device.

The samples were tested in a nitrogen atmosphere (N2), at a gas flow rate of 60 mL/min and a heat rate of 20 °C/min in the temperature range of 25–1000 °C. The analysis was performed using 10 mg of each composition.

2.7. Bending Core Cohesion

The fire behavior of the samples was determined using the method of bending core cohesion according to SR EN 520/A1:2010. The method consists of applying a bending moment to the sample while the lateral surface is affected by the flames of two burners. By heating, the bending moment causes a flexion on the sample [53]. Samples were of $300 \times 45 \times 12.5$ mm^3 size. According to the test requirements the sample needs to maintain its structure during 15min at a constant temperature of 950 °C.

2.8. Mechanical Properties

The mechanical efficacy of the volcanic rocks in the composition of the composite materials was subject to flexural and compression forces. The tests showed the capacity of the materials to react to the action of exterior forces and were operated according to SR EN 196-1:2016 at 3, 7, 14 and 28 days [54]. The flexural strength was analyzed on samples with $40 \times 40 \times 160$ ^{3} size, using an automatic flexural tensile tester L15 Controls. The samples tested for compressive strength were the remaining prisms after the flexural strength. The test was carried out with a hydraulic press of 250KN Tecnotest.

2.9. Thermal Properties

The thermal behavior of the composite based on hemp–volcanic rocks–cement was determined according to SR EN 12667:2002 [55]. The test was performed using a heat flow meter of type FOX 200 (TA Instruments). The thermal conductivity determination consisted of applying a variable heat flux to a sample measuring $150 \times 150 \times 30$ ^{3}, fitted between two plates. The report obtained after each sample test indicated the values of the thermal conductivity coefficient and thermal resistance.

2.10. Acoustic Properties

The sound absorption coefficient defines the capacity of the materials to quantify the dissipation energy of a sound. The acoustic measurements were made according to SR EN ISO 10534-2 [56], using the transfer function method with two microphones. The test was carried out with the help of the Kundt Impedance tube, on a range frequency from 50 Hz to 6400 Hz. Two types of samples, of 28 mm and 100 mm diameter, were analyzed. For the low frequency range 0 to 1600 Hz, the test was operated on the high diameter samples and for the high frequency range from 500 to 6400 Hz, on the small diameter samples. The thickness of the samples was 30 mm. According to Reference [41], the value of NRC- noise reduction coefficient can be calculated based on sound absorption coefficient. The arithmetic mean of the sound absorption coefficient on the standardized frequency ranges of 250 Hz, 500 Hz, 1000 Hz and 2000 Hz can be expressed with the following formula [57]:

$$NRC = \frac{\alpha_{250} + \alpha_{500} + \alpha_{1000} + \alpha_{2000}}{4} \ [-] \tag{1}$$

3. Results and Discussions

3.1. Chemical Characterization

The modal composition is presented in Figure 5. The X-ray diffraction testing involved the identification of crystalline compounds, characteristic to each composition. The analysis of materials based on hemp shiv, volcanic rocks and white cement shows that crystalline compounds from the initial non-hydrated phase were found inside the composition, but also new formations that appeared during the hydration process.

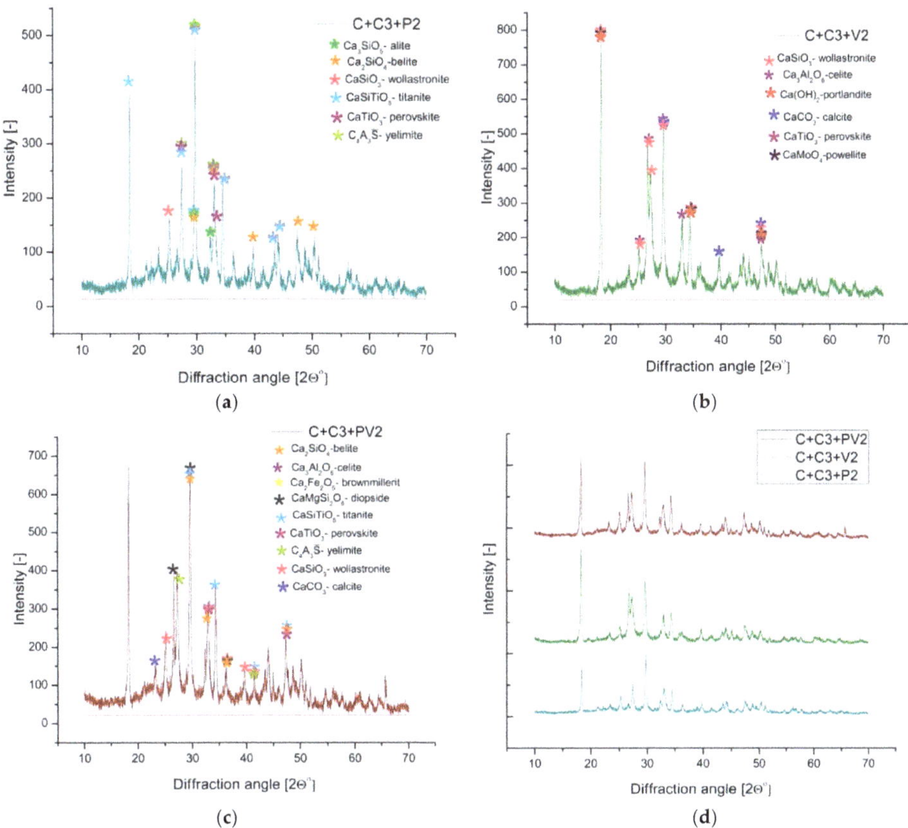

Figure 5. X-Ray diffraction (XRD) of the composite materials (**a**) M1 (C + C_3 + P_2); (**b**) M2 (C + C_3 + V_2); (**c**) M3 (C + C_3 + PV_2); (**d**) comparative analysis.

The main compounds found in the three materials were alite, belite, celite, portlandite, calcite, wollastronite, titanite, perovskite, powellite and yelimite. The first three compounds (alite, belite and celite) define the cement powder. Therefore, their identification in the composite samples can be related to the reduced amount of water during the hydration process. The ratio of water-cement used was 1:1, while the ratio of raw materials (hemp + rocks) to binder was 5:1. The explanation could be emphasized by the fact that the volume of water in the composition was not used only for the hydration of the cement paste, but was also retained by the structure of the volcanic rocks and the hemp, which has a high level of absorption. Reference [37] confirmed that vermiculite and perlite, due to their exfoliated structure, require more water, though perlite is more absorbent that vermiculite. This fact can be related to Diagram 5 (b), of the composite M2, where the compounds of alite and belite were absent, which can demonstrate that the hydration process was more complete. The cement hydration leads to new crystalline formations. From the reaction of alite and belite with water, portlandite $Ca(OH)_2$ and gels C-S-H calcium silicate hydrate are formed, and by the hydration of celite with gypsum, ettringite $3CaO \cdot Al_2O_3 \cdot 3CaSO_4 \cdot 32H_2O$ is formed [58]. This information can be confirmed with the diagram of vermiculite composite where portlandite appears. The calcium carbonate $CaCO_3$ present on the diagrams appears as a result of the aging process of the cement paste, the X-ray analysis being performed one year after the casting of the compositions. The rest of the compounds found in the cement paste structure are wollastronite, dioside, titanite, perovskite, powellite, yelimite. Their occurrence can be explained by the hydration reactions through cement compounds, hemp and volcanic rock compounds. A common compound for the three mixtures is wollastronite CaSiO3 which is responsible for increasing the mechanical performances [59]. The titanite $CaSiTiO_5$ and perovskite $CaTiO_3$ are also present due to TiO_2, contained in the white cement power and in the volcanic rocks. Diopside is a compound found in M3, shown in Figure 5c, which has a monoclinic structure pyroxene mineral with a high melting point of 1391 °C [60]. The analysis of graphs M1, shown in Figure 5a, and M2, shown in Figure 5b, shows that the crystalline formations are different, but their compounds are found in mixture M3. From Figure 5d, by analyzing the peaks of the compositions, it can be shown that the structure of the M1 recipe is the one with the best contoured peaks.

3.2. Petrographic Features

The scanning electron microscopy (SEM) images of the three mixtures are presented in Figure 6. In order to have an overview of each microscopic structure of the composites, the SEM representations were taken to various scales, of 2 mm, 200 μm, 50 μm, and 5 μm. The morphology of the composites presented at the 2 mm scale, shown in Figure 6a–c, showed a general texture in which the hemp fibers are incorporated in a binder matrix, while at the scale of 200 μm, the cracks on the surface of the composite are more visible and the images of volcanic rocks become more prevalent. The need to obtain a small piece of composite to perform this microscopic analysis determined the application of a breaking force on the structure of the composite, which could explain some of the cracks that appeared on the surface of the material. The hemp shiv has a tendency to fix itself to the cement matrix, due to the streaks given by the woody surface. Figure 6f presents the M2 composition at 200 μm, with the channels still visible, after the hemp detachment, supports the previous observation. At the same resolution, the compositions of M1, shown in Figure 6d, and M3, shown in Figure 6e, showed the porous texture of these materials, it being easy to associate the initial SEM images of the perlite, shown in Figure 2b, and vermiculite, shown in Figure 2d, with the images from the matrix of the composite materials. Increasing the scan resolution at 50 μm, (see Figure 6g–i) allows a more accurate observation of the cement matrix covering the surface of the raw materials. The most interesting images were taken from scale 5 μm, (Figure 6j–l) showing the new structures obtained after the cement hydration. The cement hydration process is defined when the mix between the cement and water decrease the plasticity and increase the rigidity in the hardening stage [59]. The resulting compounds after the hydration of the composite materials are C-S-H gels, CH (portlandite or calcium hydroxide), ettringite, monosulfatem unhydrated, cement particles, air voids, according to References [61,62]. The first three compounds are responsible for the strength properties. The C-S-H gel has the role of filling in the pore space of the composite structures and is responsible for the strength and durability of the cement-based materials, by creating a reticular network between the cement molecules. The CH structure is represented by large crystals with hexagonal prism forms [61,63,64], while the ettringite structure is similar to needle-shaped crystals [65]. The study showed [66] that the ratio of water/cement can also impact significantly the expansion of the new hydrated crystals. When the ratio is high—in our situation, it is 1:1—there is a considerable percentage of water and space in the structure of the composites, which allows the widespread development of the new crystals. At the scale of 5 μm, a large surface covered with acicular formations can be observed; different studies refer to them as ettringite crystals. The expansive surface of ettringite can be explained through Reference [65], which states that the formation of the crystals is due to the consumption of yelimite, anhydrite, gypsum, and $Ca(OH)_2$. Some of these crystalline compounds were found in the X-ray diagrams of the new composites based on cement, hemp and volcanic rocks.

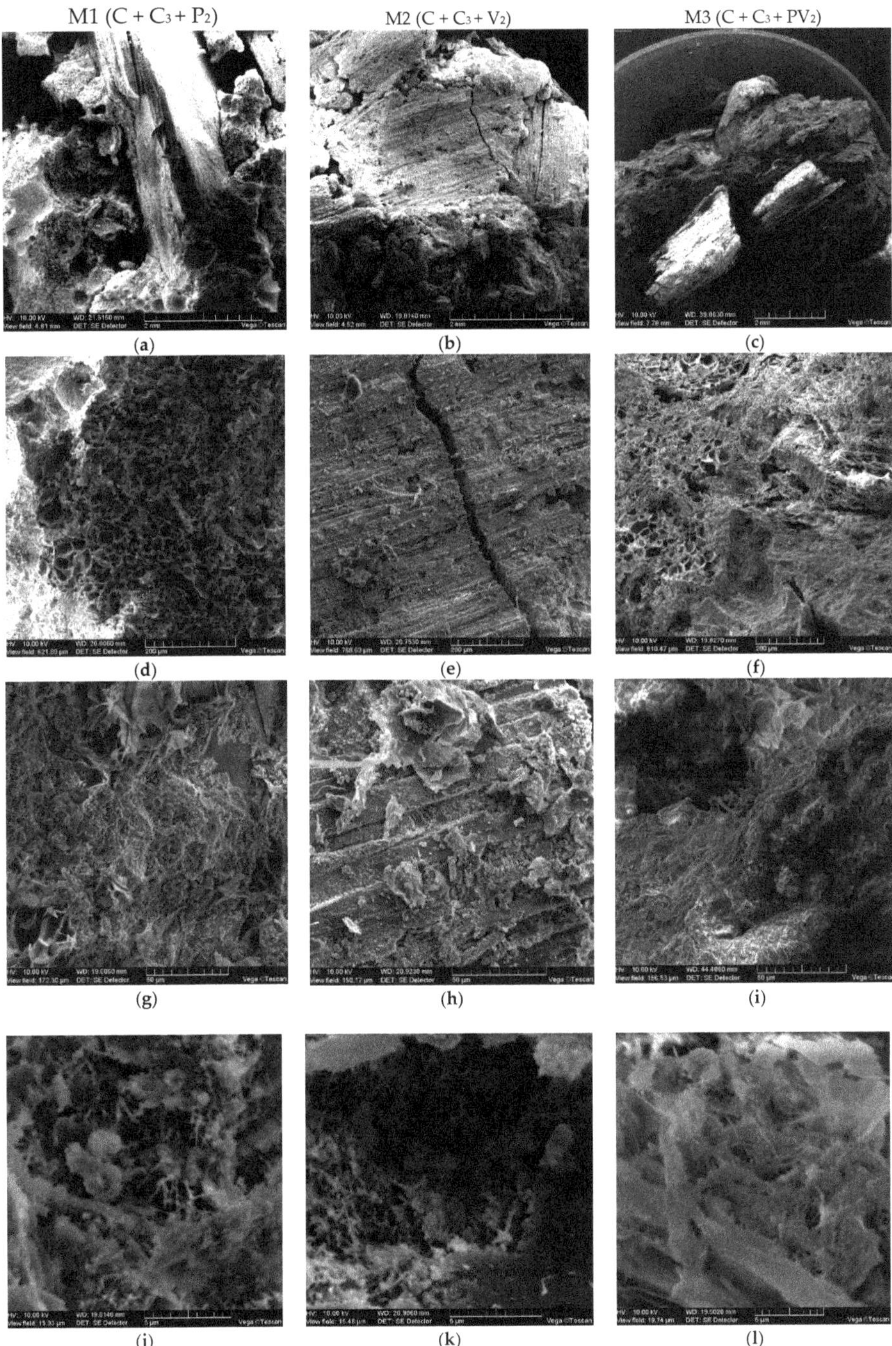

Figure 6. Scanning electron microscopy (SEM) of the composite materials at different scales at 2 mm (**a**–**c**); at 200 µm (**d**–**f**); at 50 µm (**g**–**i**) and at 5 µm (**j**–**l**).

3.3. Thermal Analysis

The thermal behavior of the composites, based on hemp shiv, volcanic rocks and white cement, is presented in Figure 7a–c. A first observation of the diagrams refers to the curves representation, which shows a degradation rate around 1% between the three composites. The lowest degradation percentage is registered by the composition M2 with 31%, followed by M3 with 32.63%, and the highest percentage being obtained by M1 with 33.24%, Figure 7d. In the degradation process, four temperature ranges were identified. Table 2 shows the amount of mass lost by each composite material, by temperature range.

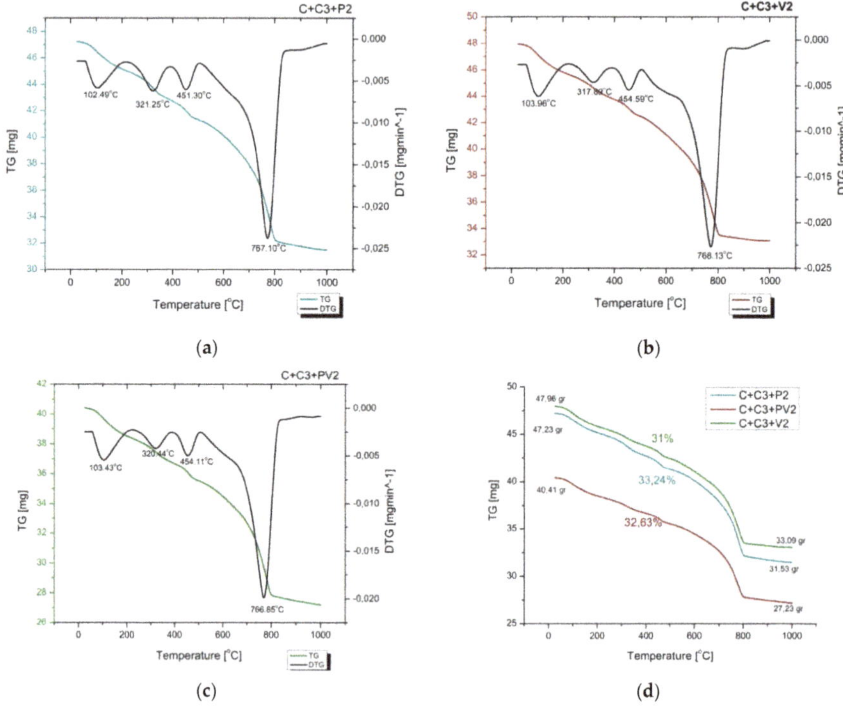

Figure 7. Thermal analysis of the composite materials (**a**) M1 (C + C$_3$ + P$_2$); (**b**) M2 (C + C$_3$ + V$_2$); (**c**) M3 (C + C$_3$ + PV$_2$); (**d**) comparative analysis.

Table 2. Mass loss of the composite.

Composition	M1 (C + C$_3$ + P$_2$)		M2 (C + C$_3$ + V$_2$)		M3 (C + C$_3$ + PV$_2$)	
UM	%	mg	%	mg	%	mg
	100	47.23	100	47.93	100	40.41
0–220 °C	4.63	2.19	4.71	2.26	4.98	2.01
220–400 °C	4.86	2.29	4.01	1.92	4.19	1.69
400–500 °C	2.94	1.39	2.69	1.29	2.86	1.15
500–1000 °C	20.80	9.82	19.06	9.40	20.57	8.31
total	33.24	15.70	31.00	14.87	32.63	13.18

The first temperature range (0–220 °C) is characterized by a loss of mass between 4.63% and 4.98% per composition, especially around the temperature of 103 °C. The highest percentage is attributed to M3. From 220 °C to 400 °C, the mass loss was similar to the previous, with an exception—M1 degraded quicker, as shown in Figure 7a. On the third interval the maximum percentage loss was 2.94% for M1, at around 450 °C. For the temperature range from 500 to 1000 °C there was approximately a 20% mass loss at the temperature of 767 °C. According to the scientific literature, the thermal analysis of the cement paste is characterized by the discomposure of ettrinigite at a temperature of 120–130 °C, the C-S-H gel below 150 °C, the non-hydrated gypsum at about 140 to 170 °C and CH (portlandite) between 420 and 550 °C [67]. Reference [28] presented the thermal behavior of the hemp shiv treated with $CaOH_2$ (lime solution) and identified two temperature ranges for the degradation, the first one around 337 °C—when the depolymerisation of hemicellulose or pectin took place—and the second one, between 377–399 °C, when the cellulose decomposition occurred.

The perlite thermal decomposition presented in Reference [68] is defined by two stages, up to 120 °C the volcanic rock loses the water, and before 600 °C the dehydroxylation process occurs. Vermiculite, according to Reference [69], loses the water molecules under 127 °C. After that, it releases the hydroxyl from the interlayers. An observation of the four temperature ranges for the above mentioned composite materials in correlation with the scientific literature is that from 0–220 °C the composite mortars lose the water molecules, which affects the decomposition of the ettringite, CSH and non-hydrated gypsum. From 220 to 400 °C, the hemp shiv is decomposed by losing cellulose and hemicellulose. From 400 to 500 °C the portlandite crystals disappear and from 500 to 1000 °C the volcanic rocks release the hydroxyl groups.

3.4. Bending Core Cohesion

The composite materials were tested for bending core cohesion and the results complied with the test requirements, as shown in Figure 8b. The samples were tested at their thickness real scale, considered as prefabricated plates. The materials were embedded in a device in which the side faces were positioned between two burners, as shown in Figure 8a. After 15 min, at a temperature of 950 °C, the three compositions remained intact. The opening produced by the flame was 7.4 cm for M1, 7 cm for M2 and 6.5 cm for M3, according to Figure 8c. Positioned vertically (see Figure 8d) the samples show the degree of fire penetration inside the material. Therefore, it can be noticed that the sample with hemp-perlite-vermiculite is more resistant due to the fact that the fire penetration is less expanded comparative with the other samples. The most damaged sample was the perlite composition.

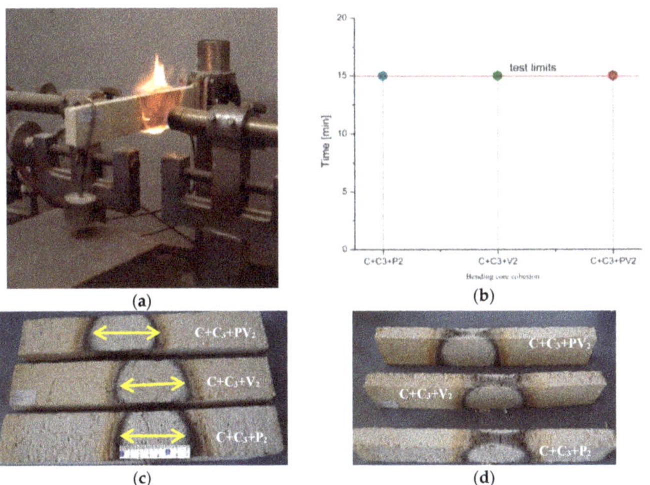

Figure 8. Bending core cohesion of the materials (**a**) testing device; (**b**) graphic results; (**c**) horizontal faces of the samples; (**d**) vertical faces of the samples

3.5. Mechanical Properties

The results of the flexural and compressive strengths are presented in Figure 9a,b. The tests were performed at 3, 7, 14 and 28 days, with the flexural strength device, shown in Figure 9c and the hydraulic press, shown in Figure 9d, for compressive strength.

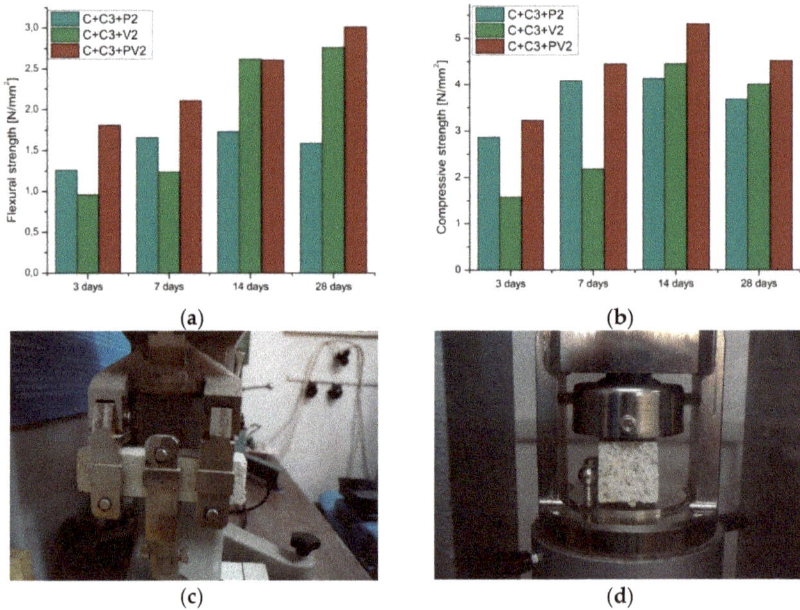

Figure 9. Mechanical properties (**a**) flexural strength; (**b**) compressive strength; (**c**) flexural strength device; (**d**) hydraulic press.

For the flexural strength, the three composites show values between 1.6–3 N/mm^2, while for the compressive strength the values are between 3.69 and 4.53 N/mm^2. As a general observation, the results obtained were lower than the standard cement mortar values, because of the hemp shiv and volcanic rocks incorporated in the matrix of the new elements. reference [37] states that the air from the pores of the volcanic rocks is responsible for the decrease in the mechanical properties of the new materials. Another aspect is that the hemp and volcanic rocks absorb the water, but do not react to it, so the only element that creates bonds and connects all raw materials is the cement [37]. There is a continuous increase of the compressive and flexural strengths during the tests. A common observation related to the mechanical resistances refers to the impact of the volcanic rocks upon the structure of the composites. M1 sample evolves the least, the values recorded being close to one another. There is a slight gradual increase at 3, 7 and 14 days, followed by a decrease at 28 days, although the difference is not particularly significant. The decreased values obtained for compressive strength at 28 days may be explained by the environmental conditions. Due to this fact the hemp-based materials were designed as prefabricated plates, and the investigated samples were kept in laboratory conditions. The porous volcanic rocks from the samples, absorbed more water, which affected the proper hydration of the cement until 28 days In contrast, M2 has a relatively small increase in values between 3 and 7 days, after which the resistance, both at bending and at compression, doubles at 14 days. At 28 days, the value of the bending resistance increases compared to 14 days value, while the compression strength values decrease. Using both perlite and vermiculite in M3 exhibits an improvement of the mechanical strengths, the flexural and compression values registering a constant increase with similar or higher values compared to the other two compositions. An explanation on the performances of perlite and vermiculite composites could be shown in the X-ray analysis, where the crystal compound of diopside was identified. According to Reference [45], diopside presented a lower degradation rate and increased the mechanical performances. The hemp shiv size can be also taken into consideration for the poor values of the mechanical strength, as References [70,71] showed that, the finer the hemp shiv particles, the stronger the boundaries around the fibers created by the binder will be, and the higher the mechanical characteristics of the composite. The values obtained for the compressive strength are similar to those in Reference [72].

3.6. Thermal Properties

The thermal properties of the composite materials are defined by bulk density, thermal conductivity and thermal resistance. The thermal conductivity and the thermal resistance were investigated using the Fox 200, Figure 10a, while for the bulk density of the samples, their mass was divided by their volume. The bulk density of the materials is between 600 and 750 kg/m^3, Figure 10b, the thermal conductivity varies between 0.124 to 0.162 W/mK, Figure 10c and thermal resistance values are between 0.189 and 0.244 m^{2}K/W, Figure 10d.

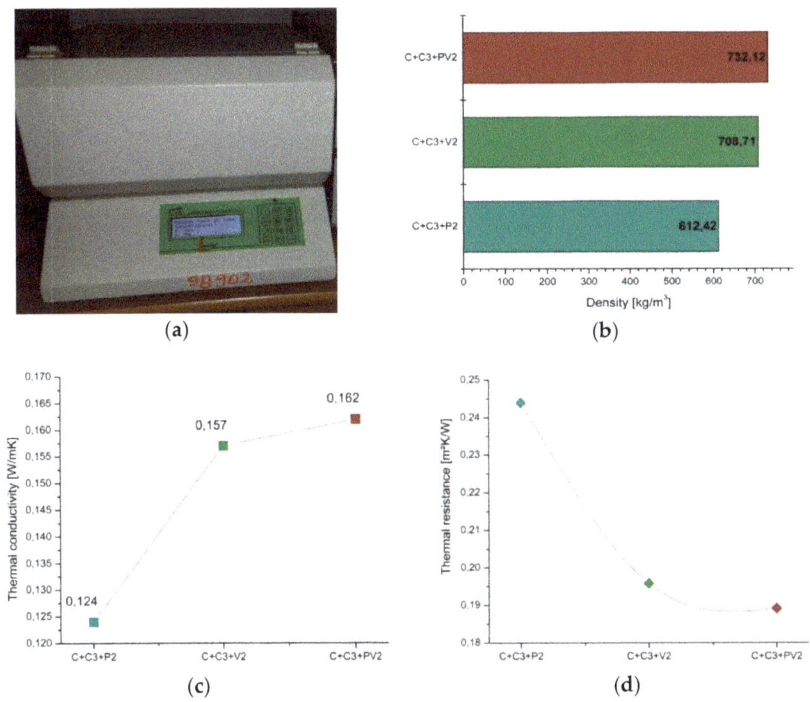

(a) (b)

(c) (d)

Figure 10. Thermal properties (**a**) device; (**b**) density; (**c**) thermal conductivity; (**d**) thermal resistance

Due to the fact that the compositions are expressed in volumes, the differences that occur are based on the type of volcanic rock used. In the pre-testing phase, the density of the two types of rocks was determined. The results show that perlite was lighter than vermiculite, based on the values obtained (130 kg/m^3 for perlite, 140 kg/m^3 for vermiculite). Therefore, the best thermal performance is recorded by M1 with 0.124 W/mK and a density of 612 kg/m^3 compared to M2. The same observation was made also in Reference [73]. The values of the thermal conductivity are directly proportional to the density of the materials, while the thermal resistance is inversely proportional to the thermal conductivity.

The M3 composition shows a lower density than that of M2, which can be related to the different particle size distribution of both fillers. Further investigation on the particle size distribution of these rocks should be carried out. The addition to volcanic rocks in hemp-cement mortar compositions results in an increase of thermal performance, to around 50% [18]. In the analysis of their mechanical strengths compared to the thermal performance of the materials one can see that the addition of perlite increases the thermal performance of a material and at the same time decreases the values of mechanical strength. A consideration regarding the improvement of the thermal properties could be connected to the mixing process, according to Reference [73]. If mixing time is increased, the volcanic rocks are broken, which will increase the density and decrease the value of the thermal conductivity. Regarding the influence of the hemp in the compositions, it can be considered that using a smaller dimension of the wood fibers will grow the thermal properties due to the fact that the materials will be defined by a finer porous structure [72].

3.7. Acoustic Properties

The acoustic properties of the new composite materials are defined by the sound absorption coefficient according to the frequency range between 0–6400 Hz; they are tested using the Kundt tube, as shown in Figure 11a.

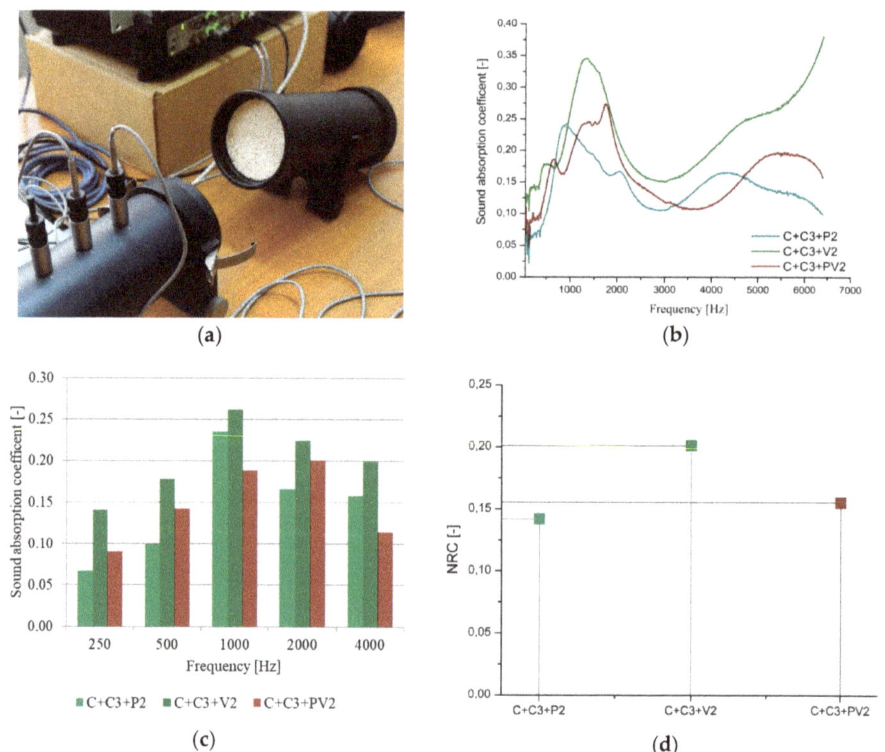

Figure 11. Acoustic properties (**a**) Kundt tube; (**b**) sound absorption coefficient 0–6400 Hz; (**c**) sound absorption coefficient on frequency standard bands; (**d**) noise reduction coefficient graph

The sound absorption coefficient is defined by the absorbed and incident sound energy [74]. The diagrams of the sound absorption coefficients on a frequency range from 0 to 6400 Hz are presented in Figure 11b. The capacity of a material to absorb the sound is defined by the type of binder, its percentage, the size of the raw materials and the thickness of the new material [75]. The data obtained after the measurements, shown in Figure 11b, showed that the maximum value of sound absorption from the three composites is around 0.35 on a frequency of around 1500 Hz, for M3. Analyzing the sound absorption coefficient on the standard frequency bands, show in Figure 11c, one can see that M1 and M2 gradually increase sound absorption up to 1000 Hz and then they reduce it. Composite M3 presents increasing values until 2000 Hz followed by almost 0.1, on the last frequency standard band Figure 11d shows the noise reduction coefficient which confirms that the most absorbent sample is defined by M2 because of its higher porous structure confirmed through the SEM images, shown in Figure 2b,d. Less absorbent are M3 and M1. This is confirmed by Reference [76], that smaller particles increase the reflection and larger particles increase the attenuation and reduce the reflection. Comparing these results with other studies [77] it was remarked that the sound absorption coefficient depends on the particle size of the hemp and the type of binder. The cement is regarded as a reflexive binder; thus hemp and volcanic rocks will have the main role in the absorption of the sound. At the same time, increasing the volume of binder, which has low mechanical properties, will negatively influence the porosity and will decrease sound absorption. The size of the hemp shiv can influence the sound absorption coefficient of the material: if the size of the hemp shiv is smaller, the sound absorption will increase [77]. Considering all the information found in the scientific literature regarding the size of the particles, the binder, the thickness of the composite, and the air gap, further research shall be carried out in order to analyze all the parameters that influence the sound absorption coefficient of the new industrial crops composites.

4. Conclusions

The new hemp shiv, volcanic rock and cement based mortars studied in this paper were designed to respond to sustainable conditions, but also to increase indoor comfort performance. The main conclusions resulting from the study are:

- The X-ray diffraction helped interpret the hydration process of the cement related to the raw materials;
- The SEM images showed new crystalline bonds that formed in the composition of the new hemp based materials;
- The thermal analysis revealed that, by adding vermiculite to the composition, the degradation of the composites at higher temperature will be slowed down;
- For the fire behavior, the presence of perlite and vermiculite increased the fire resistance of the organic material;
- The mechanical characteristics of the materials are especially defined by the ratio of the binder with water and binder with raw materials. The best performing composite for flexural and compressive strength is the one based on perlite and vermiculite;
- The thermal conductivity is related to the bulk density of the composite. Increasing the pore structure of a material will contribute to a lower density, which will lead to a decreased value of the thermal conductivity. By using perlite in the composites the value of the thermal conductivity will decrease;
- In terms of sound absorption coefficient, the presence of vermiculite in the mixtures increases the acoustic performance of the new composite materials.

Author Contributions: Conceptualization, R.I.; methodology and investigation, R.I., D.R.T.-G., D.L.M.; writing—original draft preparation, R.I.; writing—review and editing R.I., D.R.T.-G.; funding acquisition, D.R.T.-G. All authors have read and agreed to the published version of the manuscript.

Funding: The results presented in this paper were obtained in the framework of the GNaC 2018 ARUT grant "Innovative solutions for the acoustic comfort in open space offices", research Contract no. 3223/06.02.2019, with the financial support of the Technical University of Cluj-Napoca.

Conflicts of Interest: The authors declare no conflict of interest.

References

1. Tuladhar, R.; Yin, S. *Sustainability of Using Recycled Plastic Fiber in Concrete*; Elsevier Ltd.: Amsterdam, The Netherlands, 2019.
2. Maria, C.; Andreş, D.; Manea, D.L. Using Wheat Straw in Construction. Available online: https://www.google.com.hk/url?sa=t&rct=j&q=&esrc=s&source=web&cd=1&ved=2ahUKEwiu9fT668XnAhXFdd4KHSR_CwQQFjAAegQIAhAB&url=http%3A%2F%2Fjournals.usamvcluj.ro%2Findex.php%2Fpromediu%2Farticle%2Fdownload%2F11216%2F9198&usg=AOvVaw0myBXdWhQ_o3NqZBIMrblK (accessed on 10 February 2020).
3. Jami, T.; Karade, S.R.; Singh, L.P. A review of the properties of hemp concrete for green building applications. *J. Clean. Prod.* **2019**, *239*, 117852. [CrossRef]
4. Tamas-Gavrea, D.-R.; Istoan, R.; Tiuc, A.E. Multilayered Composite Panel and the Method Used for Obtaining it. Patent Application No. A00288, 24 April 2018.
5. Hamard, E.; Cazacliu, B.; Razakamanantsoa, A.; Morel, J. Cob, a vernacular earth construction process in the context of modern sustainable building. *Build. Environ.* **2016**, *106*, 103–119. [CrossRef]
6. Dénes, O.; Florea, I.; Manea, D.L.; Afonso, P. Utilization of sheep wool as a building material. *Procedia Manuf.* **2019**, *32*, 236–241. [CrossRef]
7. Petrounias, P.; Giannakopoulou, P.P.; Rogkala, A.; Lampropoulou, P.; Tsikouras, B.; Rigopoulos, I. Petrographic and Mechanical Characteristics of Concrete Produced by Different Type of Recycled Materials. *Geosciences* **2019**, *9*, 264. [CrossRef]
8. Abdallah, S.; Fan, M. Characteristics of concrete with waste glass as fine aggregate replacement. *Int. J. Eng. Technol. Res.* **2014**, *2*, 6.
9. Tiuc, A.B.; Rusu, T.; Vasile, O. Investigation Composite Materials for its Sound Absorption Properties. Available online: https://www.researchgate.net/publication/258212172_Investigation_composite_materials_for_its_sound_absorption_properties (accessed on 10 February 2020).
10. Bedlivá, H.; Isaacs, N. Hempcrete—An environmentally friendly material? *Adv. Mater. Res.* **2014**, *1041*, 83–86. [CrossRef]
11. Hamzaoui, R.; Guessasma, S.; Abahri, K. Mechanical Performance of Mortars Modified with Hemp Fibres, Shives and Milled Fly Ashes. Available online: https://www.researchgate.net/publication/319879827_Mechanical_performance_of_mortars_modified_with_hemp_fibres_shives_and_milled_fly_ashes (accessed on 10 February 2020).
12. Hirst, E.A.J. Characterisation of Hemp-Lime As a Composite Building Material. Available online: https://researchportal.bath.ac.uk/en/publications/characterisation-of-low-density-hemp-lime-composite-building-mate (accessed on 10 February 2020).
13. Mazhoud, B.; Collet, F.; Pretot, S.; Chamoin, J. Hygric and thermal properties of hemp-lime plasters. *Build. Environ.* **2016**, *96*, 206–216. [CrossRef]
14. Kinnane, O.; Reilly, A.; Grimes, J.; Pavia, S.; Walker, R. Acoustic absorption of hemp-lime construction. *Constr. Build. Mater.* **2016**, *122*, 674–682. [CrossRef]
15. Diquélou, Y.; Gourlay, E.; Arnaud, L.; Kurek, B. Impact of hemp shiv on cement setting and hardening: Influence of the extracted components from the aggregates and study of the interfaces with the inorganic matrix. *Cem. Concr. Compos.* **2015**, *55*, 112–121. [CrossRef]
16. Diquélou, Y.; Gourlay, E.; Arnaud, L.; Kurek, B. Influence of binder characteristics on the setting and hardening of hemp lightweight concrete. *Constr. Build. Mater.* **2016**, *112*, 506–517. [CrossRef]
17. Balčiūnas, G.; Pundienė, I.; Boris, R.; Kairytė, A.; Žvironaitė, J.; Gargasas, J. Long-term curing impact on properties, mineral composition and microstructure of hemp shive-cement composite. *Constr. Build. Mater.* **2018**, *188*, 326–336. [CrossRef]
18. Fernea, R.; Tămaş-Gavrea, D.R.; Manea, D.L.; Roşca, I.C.; Aciu, C.; Munteanu, C. Multicriterial Analysis of Several Acoustic Absorption Building Materials Based on Hemp. *Procedia Eng.* **2017**, *181*, 1005–1012. [CrossRef]
19. Fernea, R.; Florea, I.; Manea, D.L.; Păşcuţă, P.; Tămaş-Gavrea, D.R. X-ray diffraction study on new organic-natural building materials. *Procedia Manuf.* **2018**, *22*, 372–379. [CrossRef]

20. Koksal, F.; Gencel, O.; Kaya, M. Combined effect of silica fume and expanded vermiculite on properties of lightweight mortars at ambient and elevated temperatures. *Constr. Build. Mater.* **2015**, *88*, 175–187. [CrossRef]

21. Silva, L.M.; Ribeiro, R.A.; Labrincha, J.A.; Ferreira, V.M. Cement & Concrete Composites Role of lightweight fillers on the properties of a mixed-binder mortar. *Cem. Concr. Compos.* **2010**, *32*, 19–24.

22. Garcı, P.A.; Lanzo, M. Lightweight cement mortars: Advantages and inconveniences of expanded perlite and its influence on fresh and hardened state and durability. *Constr. Build. Mater.* **2008**, *22*, 1798–1806.

23. Dunn, V. *Collapse of Burning Buildings*, 2nd ed.; PennWell: Tulsa, OK, USA, 2010.

24. Hussain, A.; Calabria-holley, J.; Lawrence, M.; Ansell, M.P.; Jiang, Y.; Schorr, D.; Blanchet, P. Development of novel building composites based on hemp and multi- functional silica matrix. *Compos. Part B* **2019**, *156*, 266–273. [CrossRef]

25. Pantawee, S.; Sinsiri, T.; Jaturapitakkul, C.; Chindaprasirt, P. Utilization of hemp concrete using hemp shiv as coarse aggregate with aluminium sulfate [$Al_2(SO_4)_3$] and hydrated lime [$Ca(OH)_2$] treatment. *Constr. Build. Mater.* **2017**, *156*, 435–442. [CrossRef]

26. Jia, X.; Ansell, M.P.; Hussain, A.; Lawrence, M.; Jiang, Y. Physical Characterisation of Hemp Shiv: Cell Wall Structure and Porosity. In Proceedings of the 2nd International Conference on Bio-Based Building Materials & 1st Conference on ECOlogical valorisation of GRAnular and FIbrous Materials, Clermont-Ferrand, France, 21–23 June 2017.

27. Abraham, R.E.; Wong, C.S.; Puri, M. Enrichment of cellulosic waste hemp (*Cannabis sativa*) hurd into non-toxic microfibres. *Materials* **2016**, *9*, 562. [CrossRef]

28. Terpáková, E.; Kidalová, L.; Eštoková, A.; Čigášová, J.; Števulová, N. Chemical modification of hemp shives and their characterization. *Procedia Eng.* **2012**, *42*, 931–941. [CrossRef]

29. Stevulova, N.; Cigasova, J.; Estokova, A.; Terpakova, E.; Geffert, A.; Kacik, F.; Singovszka, E.; Holub, M. Properties characterization of chemically modified hemp hurds. *Materials* **2014**, *7*, 8131. [CrossRef] [PubMed]

30. Chabannes, M.; Garcia-Diaz, E.; Clerc, L.; Bénézet, J.C. Effect of curing conditions and Ca(OH)2-treated aggregates on mechanical properties of rice husk and hemp concretes using a lime-based binder. *Constr. Build. Mater.* **2016**, *102*, 821–833. [CrossRef]

31. Delannoy, G.; Marceau, S.; Glé, P.; Gourlay, E.; Guéguen-Minerbe, M.; Diafi, D.; Nour, I.; Amziane, S.; Farcas, F. Influence of binder on the multiscale properties of hemp concretes. *Eur. J. Environ. Civ. Eng.* **2019**, *23*, 609–625. [CrossRef]

32. Brzyski, P.; Barnat-Hunek, D.; Suchorab, Z.; Lagód, G. Composite materials based on hemp and flax for low-energy buildings. *Materials* **2017**, *10*, 510. [CrossRef]

33. Stevulova, N.; Estokova, A.; Cigasova, J.; Schwarzova, I.; Kacik, F.; Geffert, A. Thermal degradation of natural and treated hemp hurds under air and nitrogen atmosphere. *J. Therm. Anal. Calorim.* **2017**, *128*, 1649–1660. [CrossRef]

34. De Oliveira, A.G.; Jandorno, J.C.; da Rocha, E.B.D.; de Sousa, A.M.F.; da Silva, A.L.N. Evaluation of expanded perlite behavior in PS/Perlite composites. *Appl. Clay Sci.* **2019**, *181*, 105223. [CrossRef]

35. Wang, W.; Wang, A. *Vermiculite Nanomaterials: Structure, Properties, and Potential Applications*; Elsevier Inc.: Amsterdam, The Netherlands, 2019.

36. Jahanshahi, R.; Akhlaghinia, B. Expanded perlite: An inexpensive natural efficient heterogeneous catalyst for the green and highly accelerated solvent-free synthesis of 5-substituted-1H-tetrazoles using [bmim]N3 and nitriles. *RSC Adv.* **2015**, *5*, 104087–104094. [CrossRef]

37. Abidi, S.; Joliff, Y.; Favotto, C. Impact of perlite, vermiculite and cement on the Young modulus of a plaster composite material: Experimental, analytical and numerical approaches. *Compos. Part B* **2016**, *92*, 28–36. [CrossRef]

38. Guenanou, F.; Khelafi, H.; Aattache, A. Behavior of perlite-based mortars on physicochemical characteristics, mechanical and carbonation: Case of perlite of Hammam Boughrara. *J. Build. Eng.* **2019**, *24*, 100734. [CrossRef]

39. Tian, Y.; Tang, Y.; Li, S.; Lv, H.; Liu, P.; Jing, Q. Voigt-based swelling water model for super water absorbency of expanded perlite and sodium polyacrylate resin composite materials. *e-Polymers* **2019**, *19*, 365–368. [CrossRef]

40. Zhou, Y.; Gan, X.; Xue, H.; Han, S.; Hou, J.; Feng, K.; Wang, X. Photocatalytic Degradation of Rhodamine B by Fe_2O_3/TiO_2 Coated Expanded Perlite. *Res. Environ. Sci.* **2017**, *30*, 1961–1969.

41. Nyenhuis, J.; Drelich, J.W. Essential Micronutrient Biofortification of Sprouts Grown on Mineral Fortified Fiber Mats. *Int. J. Biol. Biomol. Agric. Food Biotechnol. Eng.* **2015**, *9*, 981–984.

42. Sutcu, M. Influence of expanded vermiculite on physical properties and thermal conductivity of clay bricks. *Ceram. Int.* **2015**, *41*, 2819–2827. [CrossRef]

43. Wen, R.; Huang, Z.; Huang, Y.; Zhang, X.; Min, X.; Fang, M.; Liu, Y.; Wu, X. Synthesis and characterization of lauric acid/expanded vermiculite as form-stabilized thermal energy storage materials. *Energy Build.* **2016**, *116*, 677–683. [CrossRef]

44. Celik, A.G.; Kilic, A.M.; Cakal, G.O. Expanded perlite aggregate characterization for use as a lightweight construction raw material. *Physicochem. Probl. Miner. Process.* **2013**, *49*, 689–700.

45. Roulia, M.; Chassapis, K.; Kapoutsis, J.A.; Kamitsos, E.I.; Savvidis, T. Influence of thermal treatment on the water release and the glassy structure of perlite. *J. Mater. Sci.* **2006**, *41*, 5870–5881. [CrossRef]

46. Nozahic, V. Vers une nouvelle démarche de conception des bétons de végétaux lignocellulosiques basée sur la compréhension et l'amélioration de l'interface liant/végétal: Application à des granulats de chenevotte et de tige de tournesol associés à un liant ponce/. Available online: https://tel.archives-ouvertes.fr/file/index/docid/822142/filename/Nozahic-2012CLF22265.pdf (accessed on 10 February 2020).

47. Arizzi, A.; Cultrone, G. Negative Effects of the Use of White Portland Cement as Additive to Aerial Lime Mortars Set at Atmospheric Conditions. Available online: https://dialnet.unirioja.es/servlet/articulo?codigo=4293986 (accessed on 10 February 2020).

48. Prasad, R.; Mahmoud, A.E.R.; Parashar, S.K.S. Enhancement of electromagnetic shielding and piezoelectric properties of White Portland cement by hydration time. *Constr. Build. Mater.* **2019**, *204*, 20–27. [CrossRef]

49. Hosseini, T.; Flores-Vivian, I.; Sobolev, K.; Kouklin, N. Concrete embedded dye-synthesized photovoltaic solar cell. *Sci. Rep.* **2013**, *3*, 2727. [CrossRef]

50. Subaşi, A.; Emiroĝlu, M. Effect of metakaolin substitution on physical, mechanical and hydration process of White Portland cement. *Constr. Build. Mater.* **2015**, *95*, 257–268. [CrossRef]

51. Sivakugan, N.; Gnanendran, C.T.; Tuladhar, R.; Civil, M.B.K. *Engineering Materials*; Cengage Learning: Boston, MA, USA, 2018.

52. Ghalibaf, M.; Doddapaneni, T.R.K.C.; Alén, R. Pyrolytic behavior of lignocellulosic-BASED polysaccharides. *J. Therm. Anal. Calorim.* **2019**, *137*, 121–131. [CrossRef]

53. SR EN 520+A1:2010. Plăci de gips-carton. Definiţii, specificaţii şi metode de încercări. Available online: http://magazin.asro.ro/ro/standard/178726 (accessed on 10 February 2020).

54. SR EN 196-1:2016. Methods of Testing Cement. Determination of Strength. Available online: https://allbeton.ru/upload/iblock/113/bs_en_196_1_1995_methods_of_testing_cement_part_1_determination_of_strength.pdf (accessed on 10 February 2020).

55. SR EN 12667:2002. Thermal Performance of Building Materials and Products-Determination of Thermal Resistance by Means of Guarded Hot Plate and Heat Flow Meter Methods-Products of High and Medium Thermal Resistance. Available online: https://shop.bsigroup.com/ProductDetail/?pid=000000000030087852 (accessed on 10 February 2020).

56. SR EN ISO 10534-2:2002. Determination of Sound Absorption Coefficient and Impedance in Impedance Tubes. Part 2: Transfer-Function Method. Available online: https://www.iso.org/obp/ui/#!iso:std:22851:en (accessed on 10 February 2020).

57. Tiuc, A.E.; Vermeşan, H.; Gabor, T.; Vasile, O. Improved Sound Absorption Properties of Polyurethane Foam Mixed with Textile Waste. *Energy Procedia* **2016**, *85*, 559–565. [CrossRef]

58. Nguyen, D.D.; Devlin, L.; Koshy, P.; Sorrell, C.C. Quantitative X-Ray Diffraction Analysis of Anhydrous and Hydrated Portland Cement–Part 1: Manual Methods. *Adv. Mater. Res.* **2015**, *1087*, 493–497. [CrossRef]

59. Gritsch, L.; Conoscenti, G.; La, V.; Nooeaid, P.; Boccaccini, A.R. Materials Science & Engineering C Polylactide-based materials science strategies to improve tissue-material interface without the use of growth factors or other biological molecules. *Mater. Sci. Eng. C* **2019**, *94*, 1083–1101.

60. Ananda, K. *Eco-Friendly Nano-Hybrid. Materials for Advanced Engineering Applications*; Apple Academic Press: Palm Bay, FL, USA, 2016.

61. Cerro-prada, E. Cement Microstructure: Fostering Cement Fostering Photocatalysis Photocatalysis. Available online: https://www.intechopen.com/books/cement-based-materials/cement-microstructure-fostering-photocatalysis (accessed on 10 February 2020).

62. Hilal, A.A. *Microstructure of Concrete, High Performance Concrete Technology and Applications*; Intechopen: London, UK, 2016. [CrossRef]

63. Gadde, H.K. Effect of Hydration and Confinement on Micro- Structure of Calcium-Silicate-Hydrate Gels. Available online: https://pdfs.semanticscholar.org/bb10/7d64b7936133f72f613241290d84ad9b9c6b.pdf (accessed on 10 February 2020).

64. Couto, C.; Darc, J.; Godoy, G.C. Microstructural and Topographic Characterization of Concrete Protected by Acrylic Paint. *Mater. Res.* **2013**, *16*, 817–823.

65. Jun, Y.; Kim, J.H.; Kim, T. Hydration of calcium sulfoaluminate-based binder incorporating red mud and silica fume. *Appl. Sci.* **2019**, *9*, 2270. [CrossRef]

66. Aïtcin, P.C. *Portland Cement*; Elsevier Ltd.: Amsterdam, The Netherlands, 2015.

67. Ogirigbo, O. Influence of Slag Composition and Temperature on the Hydration and Performance of Slag Blends in Chloride Environments Influence of Slag Composition and Temperature on the Hydration and Performance of Slag Blends in Chloride Environments Okiemute Roland O. Available online: https://www.researchgate.net/publication/305641458_Influence_of_Slag_Composition_and_Temperat.ure_on_the_Hydration_and_Performance_of_Slag_Blends_in_Chloride_Environments (accessed on 10 February 2020).

68. Kolvari, E.; Koukabi, N.; Hosseini, M.M. Perlite: A cheap natural support for immobilization of sulfonic acid as a heterogeneous solid acid catalyst for the heterocyclic multicomponent reaction. *J. Mol. Catal. A Chem.* **2015**, *397*, 68–75. [CrossRef]

69. Xu, B.; Ma, H.; Lu, Z.; Li, Z. Paraffin/expanded vermiculite composite phase change material as aggregate for developing lightweight thermal energy storage cement-based composites. *Appl. Energy* **2015**, *160*, 358–367. [CrossRef]

70. Stevulova, N.; Kidalova, L.; Cigasova, J.; Junak, J.; Sicakova, A.; Terpakova, E. Lightweight composites containing hemp hurds. *Procedia Eng.* **2013**, *65*, 69–74. [CrossRef]

71. Arnaud, L.; Gourlay, E. Experimental study of parameters influencing mechanical properties of hemp concretes. *Constr. Build. Mater.* **2012**, *28*, 50–56. [CrossRef]

72. Balčiunas, G.; Vejelis, S.; Vaitkus, S.; Kairyte, A. Physical properties and structure of composite made by using hemp hurds and different binding materials. *Procedia Eng.* **2013**, *57*, 159–166. [CrossRef]

73. Salama, A.E.; Ghanem, G.M.; Abd-Elnaby, S.F.; El-Hefnawy, A.A.; Abd-Elghaffar, M. Behavior of thermally protected RC beams strengthened with CFRP under dual effect of elevated temperature and loading. *HBRC J.* **2012**, *8*, 26–35. [CrossRef]

74. Koizumi, T.; Tsuijuchi, N.; Adachi, A. The Development of Sound Absorbing Materials Using Natural Bamboo Fibers and Their Acoustic Properties. Available online: https://www.researchgate.net/publication/285676084_The_development_of_sound_absorbing_materials_using_natural_bamboo_fibers (accessed on 10 February 2020).

75. Tiuc, A.E.; Nemeş, O.; Vermeşan, H.; Toma, A.C. New sound absorbent composite materials based on sawdust and polyurethane foam. *Compos. Part B Eng.* **2019**, *165*, 120–130. [CrossRef]

76. Ghofrani, M.; Ashori, A.; Mehrabi, R. Mechanical and acoustical properties of particleboards made with date palm branches and vermiculite. *Polym. Test.* **2017**, *60*, 153–159. [CrossRef]

77. Glé, P.; Gourdon, E.; Arnaud, L. Acoustical properties of materials made of vegetable particles with several scales of porosity. *Appl. Acoust.* **2011**, *72*, 249–259. [CrossRef]

Article

Comparison of Material Properties of SCC Concrete with Steel Fibres Related to Ingress of Chlorides

Petr Lehner [1,*]**, Petr Konečný** [1] **and Tomasz Ponikiewski** [2]

[1] Department of Structural Mechanics, Faculty of Civil Engineering, VSB-Technical University of Ostrava, Ludvíka Podéště 1875/17, 708 33 Ostrava-Poruba, Czech Republic; petr.konecny@vsb.cz

[2] Department of Building Materials and Processes Engineering, Faculty of Civil Engineering, Silesian University of Technology, Akademicka 2A, 44-100 Gliwice, Poland; Tomasz.Ponikiewski@polsl.pl

* Correspondence: petr.lehner@vsb.cz; Tel.: +420-597-321-391

Received: 23 February 2020; Accepted: 18 March 2020; Published: 20 March 2020

Abstract: The paper focuses on the evaluation of chloride ion diffusion coefficient of self-compacting concrete with steel fibre reinforcement. The reference concrete from Ordinary Portland Cement (OPC) and Self-Compacting Concrete (SCC) with several values of added steel fibres—0%, 1% and 2% of weight—were cast in order to investigate the effect of fibres. The three procedures of diffusion coefficient calculation are presented—rapid chloride penetration test, accelerated penetration tests with chloride as well as the surface measurement of electrical resistivity using Wenner probe. The resulting diffusion coefficients obtained by all methods are compared and evaluated regarding the basic mechanical properties of concrete mixtures.

Keywords: SCC-SFR; chlorides; diffusion; mechanical properties; concrete

---

1. Introduction

Self-compacting concrete (SCC) [1,2] allows the simplification of concrete processing technology and the production of complex cross-sections and structural shapes. The difference compared to standard concrete is mainly in mixture composition. The proportion of cement and small aggregate is higher, and the amount of plasticizer is significantly greater. The basic properties and advantages of self-compacting concrete are mentioned in [3–6].

Other options for improvement of the mechanical properties of concrete is the use of steel fibres [7,8]. The selection of quantity and type of fibres depends mainly on the purpose of application and is a matter of intensive research [9–11]. Due to this, there are also several special tests for steel fibre reinforced concrete (SFRC), which are codified in recommendations and national standards [12,13]. A strong emphasis on the correct description of the material properties is important [14].

The application of self-compacting concrete (SCC) [5] brings simplification of the concrete processing technology due to the lack of need for vibration. Improved properties are achieved by modification of the formula, i.e., by appropriate use of aggregate, cement content and plasticizers [1]. Another rapidly growing area of concrete technology is steel fibre reinforced self-compacting concrete (SCC-SFR), which is classified within the group of composites that combine conventional typical structural concrete and fibres [15–17].

The purpose of the use of fibres is that it eliminates one of the greatest concrete disadvantages–low tensile strength. Fibre concrete application examples can be found particularly in tunnel lining, industrial floors, foundations, or structural elements of carrier systems. The most demanding use is in the latter group, where it is also necessary to find a suitable design procedure while providing enough reliability and safety of the structure. These may be for instance cases of lightweight structural elements where fibres are to eliminate shear destruction/failure. Input parameters of the calculation, fibre concrete properties are verified mainly by laboratory testing included in suggestions, standards, and design code [12,13]. It is well known that electrical properties are influenced by conductive materials. Thus, it is interesting to record the electrical conductivity values of SCC-SFR mixtures.

Furthermore, the development of tools for predicting the durability of concrete structures that help to develop durable concrete and construction systems is useful. Increased focus on higher durability helps maintain the required level of safety and maintenance for a longer period of time, which saves the cost related to premature repair and reconstruction [18,19]. A correct description of the properties of the composite materials can be combined with modelling of structures within several available models [20–23].

The article aim is to determine and extend the knowledge about the properties of self-compacting concrete with steel fibre reinforcement mixture with a fibre content of 0%, 1%, and 2%. This is due to their use in the field of reinforced concrete structures exposed to chlorides, where it is necessary to properly model diffusion processes [24–27]. Moreover, three methods for evaluation of the diffusion coefficient related to a concrete's ability to resist chloride ingress are compared—electrical resistance measurements [28], the rapid chloride permeability test [29], and accelerated penetration tests with chloride [30].

2. Mixtures Properties and Samples Settings

The evaluated experiment, calculations and results are part of the extensive campaign dealing with the durability of concrete structures. The laboratory experiments were conducted at the laboratories of Silesian University of Gliwice and VSB - Technical University of Ostrava. The set of laboratory samples consisted of seven large cylinders (diameter 150 mm, height 300 mm), six smaller cylinders (diameter 100 mm, height 250, 200 and 100 mm), four cubes (dimension 150 mm), and three beams (150 × 150 mm, length 450 mm) for each mixture. (see Figure 1a).

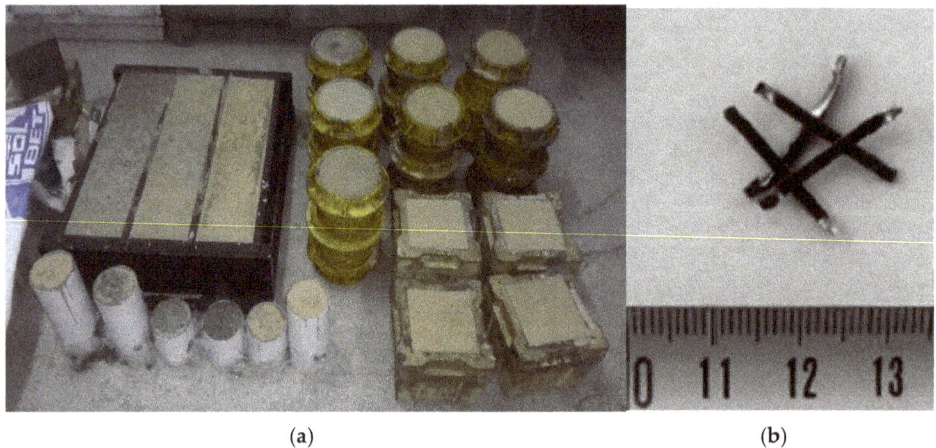

(a)   (b)

Figure 1. (a) Example of preparation of laboratory samples for one set of mixtures; (b) Steel fibres KE20/1.7—shape and dimension [31].

The reference concrete was formed from Ordinary Portland Cement (OPC). The Self-Compacting Concrete (SCC) with several values of added steel fibres—0%, 1%, and 2%—was used in order to investigate the effect of fibres. The steel fibres were of type KE20/1.7 (see Figure 1b). The composition of the mixtures is shown in Table 1, and it is based on earlier SCC research at the SUT in Gliwice [1,19,31,32]. It needs to be noted that the cement applied in the SCC mixture had been in laboratory storage for more than two years, and partial hydration had already started. However, the concrete slump test was executed, and all mixtures have the same value of workability.

Table 1. The concrete mixtures components [31] based on from Ordinary Portland Cement (OPC) and Self-Compacting Concrete (SCC) with steel fibres.

Mixture No.	OPC	SCC 0%	SCC 1%	SCC 2%
Cement type I 42.5 R	313 kg/m^3		490 kg/m^3	
Water	164 kg/m^3		201 kg/m^3	
Sand	387 kg/m^3		807 kg/m^3	
River gravel	1546 kg/m^3		807 kg/m^3	
Superplastificator	-		12.25 kg/m^3	
Stabilizator	-		1.96 kg/m^3	
Steel Fibres	-	-	80 kg/m^3	160 kg/m^3
Water/cement ratio (W/C)	0.52		0.41	

3. Experimental Tests Range and Results

A comprehensive range of tests comprised from basic mechanical properties; fracture test, as well as electrical resistance measurements [28], the rapid chloride permeability test [29], and accelerated penetration tests with chloride [30] were conducted.

3.1. Mechanical Properties

Compressive strength measurements were performed on standard cubes and cylinders, and also the tensile strength and modulus of elasticity were determined (see Figure 2) [31]. Firstly, it is possible to compare the results between an OCP and SCC mixture. In all properties, the SCC values are higher than the OPC. Because the SCC mixtures are not prepared in the way typical for high-performance concrete, it was expected that there would not be much difference in mechanical properties.

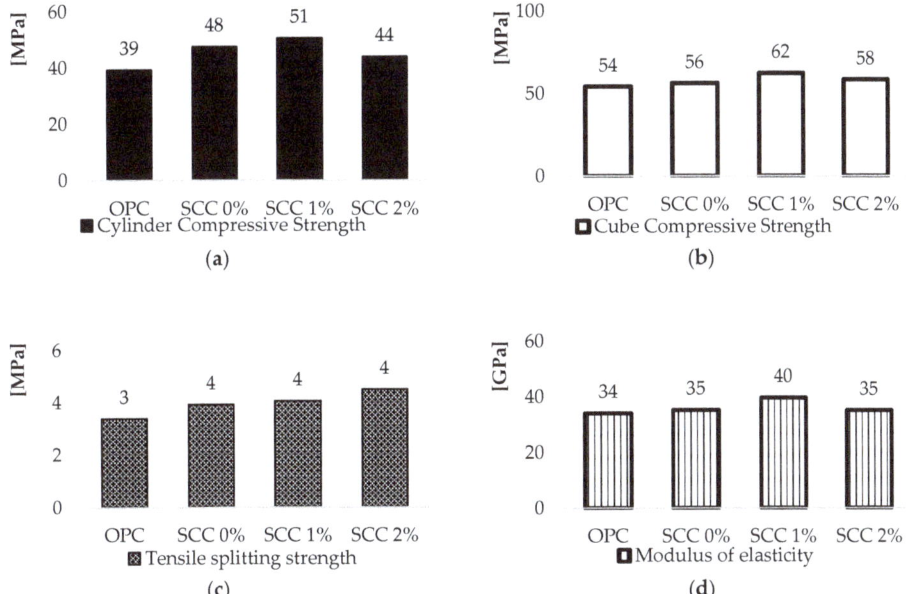

Figure 2. Results of material characteristics for all mixtures: (**a**) cylinder compressive strength; (**b**) cube compressive strength; (**c**) tensile spitting strength; (**d**) modulus of elasticity [31].

The second possibility is comparison related to the number of steel fibres in the SCC mixture. It shows that the higher the relative weight of the steel fibres, the higher the tensile splitting strength.

However, other mechanical properties show that, when the amount of fibres exceeds a certain value, the properties no longer improve. The modulus of elasticity of the mixture SCC 1% is about 15% higher compared with other mixtures. This is in line with the typical limitation of the use of steel fibres up to 1.5% of the weight of concrete [33].

3.2. Diffusion Properties

The three procedures of detection of concrete diffusion coefficient were used. Two indirect electrochemical methods—rapid chloride permeability test (RCPT) [29] and surface measurement of electrical resistivity using Wenner probe (Resistivity) [28]—and a method based on the evaluation of chloride action—accelerated penetration tests with chloride (NTBuild 443) [30].

3.2.1. Rapid Chloride Permeability Test

The RCPT test [29] was conducted at the SUT in Gliwice. Due to the influence of fibre scattering, one cylindrical core for each mixture was cut into three slices of testing and marked Upper, Middle and Lower. All three slices from one cylinder were analysed. Test specimens were prepared and test procedures were performed according to ASTM C1202 [29]. The resulting passing charges, which indirectly evaluated the ability of the concrete to withstand an aggressive environment, are shown in Figure 3. The results show that there is a negligible difference in the reference mixture in terms of individual slices. There are some deviations (approx. 15%) in SCC mixtures without wires, which can be influenced by shrinkage of the mixture. For a mixture of SCC with 1% of fibres, the differences are more significant, and the most significant differences are found for a mixture of 2% of fibres. The problem of the test is a rapid increase in temperature when using fibre reinforced concrete. From this point of view, the test is not an ideal solution for such an SCC-SFR mixture. Taking statistical variance into consideration, the average value was chosen for derivation of diffusion coefficients.

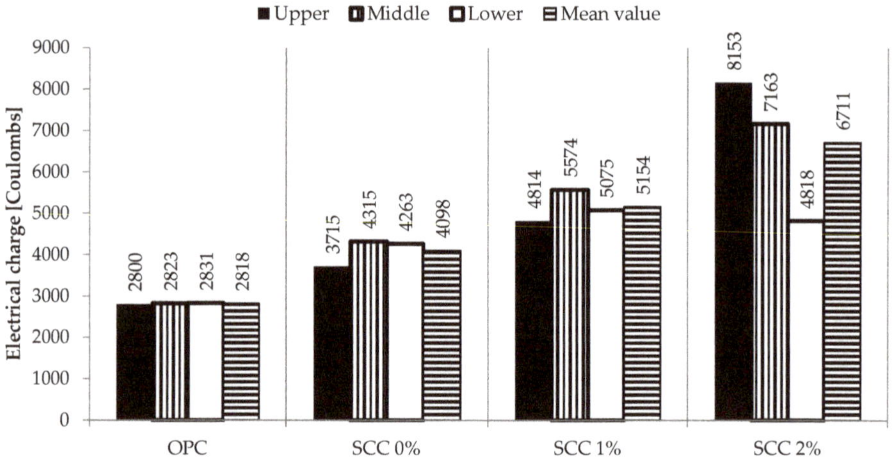

Figure 3. Results of the rapid chloride permeability test for three slices of testing and marked Upper, Middle and Lower, and mean values.

3.2.2. Electrical Resistivity

The standard test method for surface resistivity of concrete [28] is non-destructive; therefore, repeated measurements are possible to determine the time dependency of the diffusion. Unfortunately, this measurement method may have a relatively large variability, partly due to the heterogeneity of concrete and also due to the use of relatively uncontrollable contact conditions. It should be noted that the values of the volumetric resistivity were calculated from surface resistivity based on the relationship obtained from [34] and after that, to the resulting passing charges (equivalent to the results from RCPT) [35]. The test is non-destructive, and it is possible to measure the electrical properties during concrete ageing (see Figure 4). Results were partly published in [19,31].

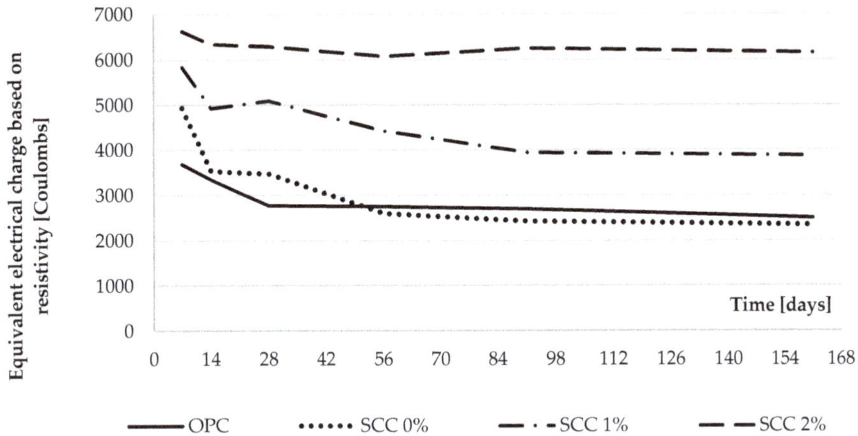

Figure 4. Results of time-dependent surface electrical resistivity of concrete.

Looking at the change of the passed charge, we may derive the following findings. The OPC mixture has standard behaviour whereby charge decreases over the concrete maturation, thus reflecting an improved ability of the concrete to withstand the aggressive environment. The SCC mixtures have similar behaviour, reflected in the similar shape of the curves. Although the relationship of shapes between 14 and 28 days are visible in all three SCC mixtures, their absolute difference cannot be considered as constant. If the influence of passed charge is based on the amount of added steel fibres, the shape of curves is proportional. However, there are also other influences that affect the readings, such as the initiation of corrosion of steel fibres.

3.2.3. NTBuild 443

The third test method was based on the modified NORDTEST NT Build 443 [13]. Concrete specimens were immersed in the saline solution (see Figure 5a), and then sampling of concrete powder in respective layers of chloride profile was conducted by drilling (see Figure 5b), thus the suitable period for chloride penetration was selected as 90 days. The concrete powder was subsequently evaluated for the presence of chloride ions in the laboratory, the obtained chloride profile was analysed, and the diffusion coefficient was calculated. The whole test procedure and the process of calculating the diffusion parameter are described in detail in [36].

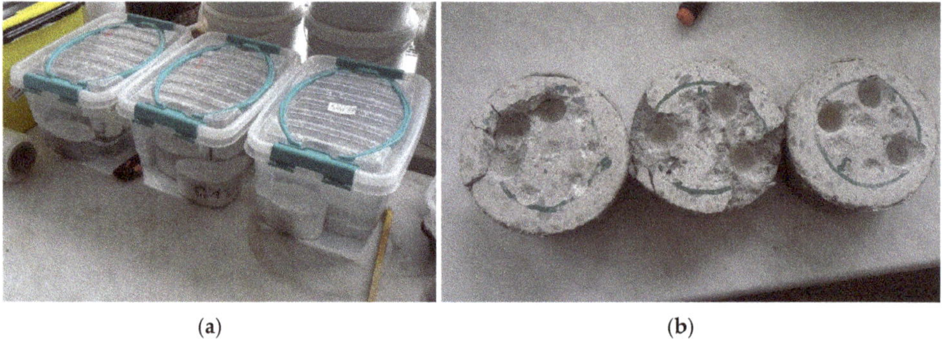

(a)   (b)

Figure 5. The process of the modified NORDTEST NT Build 443: (**a**) samples submerged in the chloride solution; (**b**) samples after drilling.

4. Comparison of Diffusion Coefficients

The resulting values of diffusion coefficients calculated from direct chloride profiling, and indirect electrochemical methods (RCPT and resistivity), are given in Figure 6 and in Table 2. The calculation of the diffusion coefficient $D_{c(t)}$ is based on the procedures given in [35,37] and is very well described in [36,38,39]. The number of analysed samples mean value and standard deviation is given in Table 2. Figure 6 shows mean value and T-plot of minimum and maximum values.

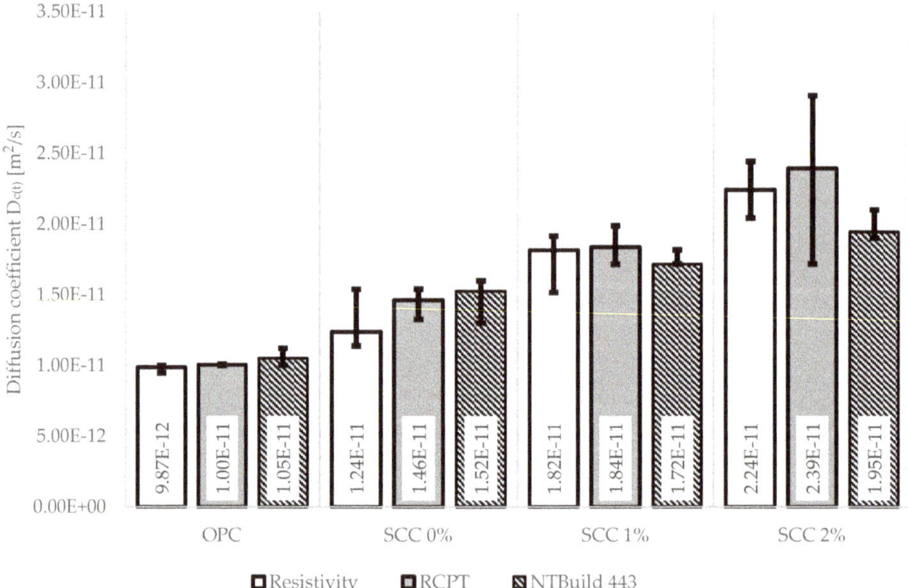

Figure 6. Diffusion coefficient $D_{c(t)}$ derived from surface resistivity, rapid chloride permeability test (RCPT) and chloride profile analysis (NTBuild 443). The minimum and maximum values are shown in the form of a T-graph.

Table 2. Statistical data of diffusion coefficient $D_c(t)$ derived from surface resistivity, rapid chloride permeability test (RCPT) and chloride profile analysis (NTBuild 443).

Mixture No.	Method	Mean Value [m²/s]	Standard Deviation [m²/s]	Number of Samples [-]
OPC	Resistivity	9.80×10^{-12}	2.76×10^{-13}	4
	RCPT	1.00×10^{-11}	5.55×10^{-14}	3
	NTBuild 443	1.05×10^{-11}	6.00×10^{-13}	3
SCC 0%	Resistivity	1.24×10^{-11}	2.08×10^{-12}	4
	RCPT	1.46×10^{-11}	1.08×10^{-12}	3
	NTBuild 443	1.52×10^{-11}	1.51×10^{-12}	3
SCC 1%	Resistivity	1.82×10^{-11}	2.08×10^{-12}	4
	RCPT	1.84×10^{-11}	1.36×10^{-12}	3
	NTBuild 443	1.72×10^{-11}	5.09×10^{-13}	3
SCC 2%	Resistivity	2.24×10^{-11}	2.00×10^{-12}	4
	RCPT	2.39×10^{-11}	5.96×10^{-12}	3
	NTBuild 443	1.95×10^{-11}	1.01×10^{-12}	3

It should be noted that the values of the diffusion coefficient from resistivity are based on measurement at 28 days, from RCPT are based on measurement at 56 days, and from chloride profile, approximately 118 days. All these values are precisely determined according to the standards. It is worth noticing that the different level of maturation at the time of testing each method does not affect the comparison between specific mixture design that matters.

It is necessary to explain the meaning of the diffusion coefficient value. A diffusion coefficient closer to zero shows better diffusion resistance and hence better resistance to chloride ion penetration. Looking at the results of diffusion coefficients (Figure 6), the reference mixture (OPC) reports almost the same values for all methods. On the other hand, SCC 0% mixture has larger differences, but is still within limits of the inaccuracies of each method. In this case, the results of SCC 0% are worse than OPC, which is probably influenced by chemical additives. This is also because it is not a mixture that has been prepared as high resistance to the chloride. This is observed even though the mechanical properties (see Figure 2) are better for SCC mixtures.

Subsequent evaluation of mixtures with 1% and 2% fibres is interesting in terms of two hypothetical effects. The first effect is related to the electrical conductivity of steel wires in concrete. This effect should be reflected in two methods—RCPT and surface resistivity. The second effect is related to the real acceleration of diffusion along the fibres at the transition level with other material with respect to a possible influence of the microscopic void structure and porosity [6,40]. This should affect all three methods. Considering the possible scatter of measurement, it is possible to evaluate the effect of the amount of fibres on the two mentioned effects accordingly. For the purpose of comparison, SCC 0% mixture results were considered as the base values (0%), and differences against the SCC 1% and 2% mixtures are shown in Figure 7. It is worth mentioning that the results of measurement of the RCPT is based on three cut out samples from one concrete cylinder. It can be seen, that the results are consistent within the OPC, SCC 0%, 1%. However, the scatter for the 2% fibre content shows much higher scatter, indicating that this amount of steel fibres is too high. The finding is consistent with recommendation that the highest applicable amount of fibres is 1.5%. It seems that the steel fibres were not spread out uniformly thought the cylinder.

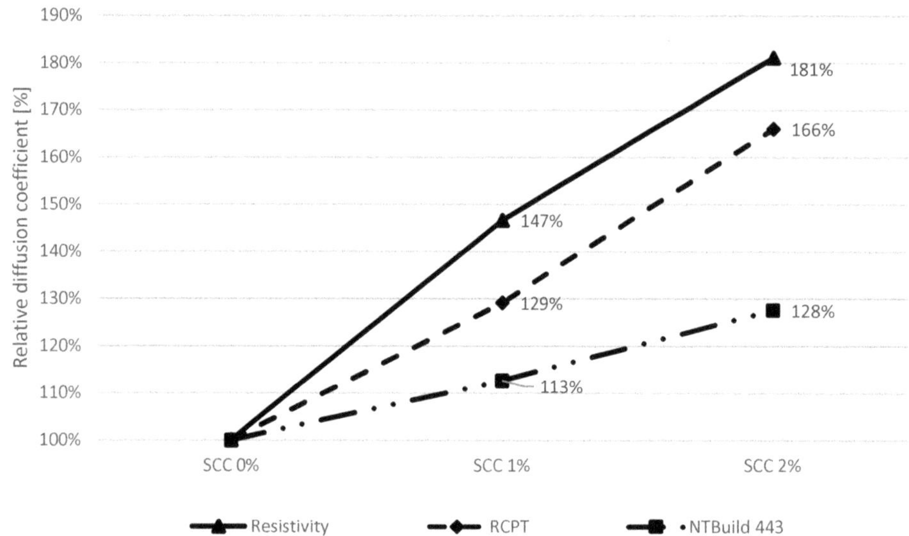

Figure 7. Percentile differences of the diffusion coefficient of the SCC 1% and 2% mixtures compared to mixture SCC 0%.

Looking at the percentage results, the resistivity, RCPT and NTBuild 443 are noticeably affected by the amount of steel fibres as expected. The most significant influence is observed with a resistivity test, which is based on the direct current measurement, and the steel material is, therefore, a significant factor. The RCPT method is less affected but is also based on the charge and use of the penetration of a salt solution (in all samples of the same amount), so the effect of the amount of the steel fibres is lower. In the third method, the evaluation of the chloride profile, it is possible to consider only the effect of increased diffusion along the fibres or the possible influence of interface and micro-cracks along the fibres.

5. Discussion and Conclusions

The article evaluates durability related to concrete material parameters. There are relative values for cube and cylinder compression strength, tensile splitting strength, and modulus of elasticity. Three approaches are discussed for the computation of diffusion coefficient applicable to the numerical modelling of chloride ion ingress to concrete. Studied approaches are chloride profiling, electrical resistivity measurement, and rapid chloride permeability test. The comparison of the relatively fast method (resistivity and RCPT) for the evaluation of concrete ability to resist aggressive agents was conducted on the sample of self-compacting concrete. There was a correlation between the amount of steel fibres and conductivity, as expected. A larger amount of the steel fibres increases the calculated diffusion coefficient, both in electrochemical methods and in chloride profile evaluation. From this point of view, it is necessary to investigate, for example, porosity in further research. It needs to be proved if it causes some worse mechanical properties and, on the contrary, may tend to increase the results of all diffusion values (reducing the resistance against chloride penetration).

From the point of view of the amount of fibres in the concrete, it would be advisable to prepare more a graduated set of mixtures, e.g., 1.2%, 1.4%, that can lead to finding the threshold at which the increasing properties change to decreasing. The experiment also confirmed that diffusion coefficients obtained by the electrochemical approaches are influenced by the steel fibres from both standpoints—electrical conductivity, as well as pore structure.

Author Contributions: Conceptualization, P.K. and T.P.; methodology, T.P., P.L. and P.K.; software, P.L.; validation, P.K., P.L. and T.P.; data curation, P.L.; writing—original draft preparation, P.L.; writing—review and editing, P.K.; visualization, P.L.; project administration, P.K. All authors have read and agreed to the published version of the manuscript.

Funding: Financial support from VSB-Technical University of Ostrava by means of the Czech Ministry of Education, Youth and Sports through the Institutional support for conceptual development of science, research and innovations for the year 2020 is gratefully acknowledged.

Conflicts of Interest: The authors declare no conflict of interest.

References

1. Pająk, M.; Ponikiewski, T. Flexural behavior of self-compacting concrete reinforced with different types of steel fibers. *Constr. Build. Mater.* **2013**, *47*, 397–408. [CrossRef]
2. Gołaszewski, J. Influence of cement properties on new generation superplasticizers performance. *Constr. Build. Mater.* **2012**, *35*, 586–596. [CrossRef]
3. Goodier, C.I. Development of self-compacting concrete. *Proc. Inst. Civ. Eng. Struct. Build.* **2003**, *156*, 405–414. [CrossRef]
4. Zhu, W.; Bartos, P.J.M. Uniformity of in situ properties of self-compacting concrete in full-scale structural elements. *Cem. Concr. Res.* **2001**, *23*, 57–64. [CrossRef]
5. Okamura, H.; Ozawa, K. Self-Compacting High Performance Concrete. *Struct. Eng. Int. J. Int. Assoc. Bridg. Struct. Eng.* **1996**, *6*, 269–270. [CrossRef]
6. Łaźniewska-Piekarczyk, B. Effect of viscosity type modifying admixture on porosity, compressive strength and water penetration of high performance self-compacting concrete. *Constr. Build. Mater.* **2013**, *48*, 1035–1044. [CrossRef]
7. Abrishambaf, A.; Barros, J.A.O.; Cunha, V.M.C.F. Tensile stress–crack width law for steel fibre reinforced self-compacting concrete obtained from indirect (splitting) tensile tests. *Cem. Concr. Compos.* **2015**, *57*, 153–165. [CrossRef]
8. Mo, K.H.; Yap, K.K.Q.; Alengaram, U.J.; Jumaat, M.Z. The effect of steel fibres on the enhancement of flexural and compressive toughness and fracture characteristics of oil palm shell concrete. *Constr. Build. Mater.* **2014**, *55*, 20–28. [CrossRef]
9. Ponikiewski, T.; Katzer, J. Fresh Mix Characteristics of Self- Compacting Concrete Reinforced by Fibre. *Period. Polytech. Civ. Eng.* **2017**, *61*, 226–231. [CrossRef]
10. Mogaveera, G.; Umesh, S.S. Anand, Research on the Strength parameters of poly propylene fiber reinforced concrete and steel fiber reinforced concrete, V.R. *Int. J. Recent Technol. Eng.* **2019**, *8*, 954–957.
11. da Silva, G.C.S.; Christ, R.; Pacheco, F.; de Souza, C.F.N.; Gil, A.M.; Tutikian, B.F. Evaluating steel fiber-reinforced self-consolidating concrete performance. *Struct. Concr.* **2019**. [CrossRef]
12. RilemT.C. 162-TDF. Test and design methods for steel fibre reinforced concrete - Bending test. *Mater. Struct. Constr.* **2002**, *35*, 579–582. [CrossRef]
13. *BS EN 14651:2005 Test Method for Metallic Fibre. Concrete—Measuring the Flexural Tensile Strength*; British Standards Institution: London, UK, 2005.
14. Katzer, J.; Domski, J. Quality and mechanical properties of engineered steel fibres used as reinforcement for concrete. *Constr. Build. Mater.* **2012**, *34*, 243–248. [CrossRef]
15. Alberti, M.G.; Enfedaque, A.; Gálvez, J.C. Fibre reinforced concrete with a combination of polyolefin and steel-hooked fibres. *Compos. Struct.* **2017**, *171*, 317–325. [CrossRef]
16. Ahmad, H.; Hashim, M.H.M.; Hamzah, S.H.; Bakar, A.A. Steel Fibre Reinforced Self-Compacting Concrete (SFRSC) performance in slab application: A review. *AIP Conf. Proc.* **2016**, *1774*, 030024.
17. Kayali, O.; Ahmed, M.S. Assessment of high volume replacement fly ash concrete-concept of performance index. *Constr. Build. Mater.* **2013**, *39*, 71–76. [CrossRef]
18. Ghosh, P.; Hammond, A.; Tikalsky, P.J. Prediction of Equivalent Steady State Chloride Diffusion Coefficients. *ACI Mater. J.* **2011**, *108*, 88–94.
19. Lehner, P.; Konecny, P.; Ponikiewski, T. Experimental and Numerical Evaluation of SCC Concrete Durability Related to Ingress of Chlorides. *AIP Conf. Proc.* **2018**, *176*, 150012.
20. Ghosh, P.; Konečný, P.; Lehner, P.; Tikalsky, P.J.P.J. Probabilistic time-dependent sensitivity analysis of HPC bridge deck exposed to chlorides. *Comput. Concr.* **2017**, *19*, 305–313. [CrossRef]

21. Novák, D.; Vořechovský, M.; Teplý, B. FReET: Software for the statistical and reliability analysis of engineering problems and FReET-D: Degradation module. *Adv. Eng. Softw.* **2014**, *7*, 179–192. [CrossRef]
22. Vořechovská, D.; Teplý, B.; Šomodíková, M. Chloride Ion Ingress Modelling and the Reliability of Concrete Structures. In Proceedings of the 12th International Conference on Structural Safety and Reliability, Vienna, Austria, 6–10 August 2017; pp. 6–10.
23. Bentz, E.C.; Thomas, M.D.A. Life-365 Service Life Prediction Model: And Computer Program for Predicting the Service Life and Life-Cycle Cost of Reinforced Concrete Exposed to Chlorides. *Life-365 User Man.* **2013**, 1–87.
24. Parant, E.; Pierre, R.; le Maou, F. Durability of a multiscale fibre reinforced cement composite in aggressive environment under service load. *Cem. Concr. Res.* **2007**, *37*, 1106–1114. [CrossRef]
25. Berrocal, C.G.; Lundgren, K.; Löfgren, I. Corrosion of Steel Bars Embedded in Fibre Reinforced Concrete under Chloride Attack: State-of-the-Art. *Cem. Concr. Res.* **2016**, *80*, 69–85. [CrossRef]
26. Azarsa, P.; Gupta, R. Electrical resistivity of concrete for. durability evaluation: A review. *Adv. Mater. Sci. Eng.* **2017**, *2017*, 1–30. [CrossRef]
27. Berrocal, C.; Lundgren, K.; Löfgren, I. Influence of Steel Fibres on Corrosion of Reinforcement in Concrete in Chloride Environments: A Review. *Fibre Concr.* **2013**, *80*, 1–10.
28. *AASHTO TP95, Standard Method of Test for Surface Resistivity Indication of Concrete's Ability to Resist Chloride Ion Penetration*; American Society for Testing and Materials: Washington, DC, USA, 2014; p. 10.
29. *ASTM C1202*; American Society for Testing and Materials: Philadelphia, PA, USA, 2012; pp. 1–8.
30. Nordtest NTBuild 443. *Nordtest Method: Accelerated Chloride Penetration into Hardened Concrete*; Nordtest: Esbo, Finland, 1995.
31. Lehner, P.; Konečnỳ, P.; Ponikiewski, T. Relationship between Mechanical Properties and Conductivity of SCC Mixtures with Steel Fibres. In Proceedings of the Central European Civil Engineering Meeting, Research and Modelling in Civil Engineering 2018, Koszalin, Poland, 4–8 June 2018.
32. Sucharda, O.; Pajak, M.; Ponikiewski, T.; Konecny, P. Identification of mechanical and fracture properties of self-compacting concrete beams with different types of steel fibres using inverse analysis. *Constr. Build. Mater.* **2017**, *138*, 263–275. [CrossRef]
33. Manoharan, S.V.; Anandan, S. Steel fibre reinforcing characteristics on the size reduction of fly ash based concrete. *Adv. Civ. Eng.* **2014**, *2014*, 1–11. [CrossRef]
34. Morris, W.; Moreno, E.I.; Sagüés, A.A. Practical evaluation of resistivity of concrete in test cylinders using Wenner array probe. *Cem. Concr. Res.* **1996**, *26*, 1779–1787. [CrossRef]
35. Ghosh, P. Computation of Diffusion Coefficients and Prediction of Corrosion Initiation in Concrete Structures. Ph.D. Thesis, The University of Utah, Salt Lake City, UT, USA, 2011.
36. Lehner, P.; Turicová, M.; Konečnỳ, P. Comparison of selected methods for measurement of the concrete electrical resistance to chloride penetration. *ARPN J. Eng. Appl. Sci.* **2017**, *12*, 937–944.
37. Lu, X. Application of the Nernst-Einstein equation to concrete. *Cem. Concr. Res.* **1997**, *27*, 293–302. [CrossRef]
38. Lehner, P.; Ghosh, P.; Konečný, P. Statistical analysis of time dependent variation of diffusion coefficient for various binary and ternary based concrete mixtures. *Constr. Build. Mater.* **2018**, *183*, 75–87. [CrossRef]
39. Konečný, P.; Lehner, P.; Ponikiewski, T.; Miera, P. Comparison of Chloride Diffusion Coefficient Evaluation Based on Electrochemical Methods. *Procedia Eng.* **2017**, *190*, 193–198. [CrossRef]
40. Roubin, E.; Colliat, J.B.; Benkemoun, N. Meso-scale modeling of concrete: A morphological description based on excursion sets of Random Fields. *Comput. Mater. Sci.* **2015**, *102*, 183–195. [CrossRef]

Article

Experimental Study on Unconfined Compression Strength of Polypropylene Fiber Reinforced Composite Cemented Clay

Qiangqiang Cheng [1,2], Jixiong Zhang [2,*], Nan Zhou [2], Yu Guo [1] and Shining Pan [1]

[1] School of Architectural Construction, Institute of Applied Technology of Construction Industrialization and Information, Jiangsu Vocational Institute of Architectural Technology, Xuzhou 221116, China; qiangcheng@cumt.edu.cn (Q.C.); hnnygy@126.com (Y.G.); panshining@163.com (S.P.)

[2] State Key Laboratory of Coal Resources and Safe Mining, China University of Mining & Technology, Xuzhou 221116, China; zhounanyou@126.com

* Correspondence: zjxiong@163.com

Received: 3 March 2020; Accepted: 23 March 2020; Published: 26 March 2020

Abstract: The effects of three main factors, including polypropylene fiber content, composite cement content and curing time on the unconfined compressive strength of fiber-reinforced cemented clay were studied through a series of unconfined compressive strength tests. The experimental results show that the incorporation of fibers can increase the compressive strength and residual strength of cement-reinforced clay as well as the corresponding axial strain when the stress peak is reached compared with cement-reinforced clay. The compressive strength of fiber-reinforced cement clay decreases first, then increases with small-composite cement at curing time 14 d and 28 d. However, fiber-reinforced cement clay's strength increases with the increase of fiber content for heavy-composite cement. The compressive strength of fiber-composite cement-reinforced marine clay increases with the increase of curing time and composite cement content. The growth rate increases with the increase of curing time. The failure mode of composite cement-reinforced clay is brittle failure, while the failure mode of fiber-reinforced cemented clay is plastic failure.

Keywords: fiber-reinforced cemented clay; unconfined compression strength; fiber content; composite cement content; curing time

---

1. Introduction

Marine clays have the characteristics of large pore ratios, high water content, low shear strength, soft sensitive and high compressibility [1,2]. Soft clays are widely present in offshore areas and cannot be used directly in geotechnical engineering activities, such as subgrade engineering, embankments, deep excavation and underground construction. Considering the economy and effectiveness, using cement to reinforce soft clay is of great popularity [3–8], compared with other chemical stabilization methods. Though cement-stabilized clay has the advantages of rapid formation, good plasticity and high compressive strength, it also has the disadvantages of low tensile strength and flexural strength. A number of studies that used fly ash, a by-product of coal or solid waste, to partially replace cement for improving the mechanical strength of cement-stabilized clay have been carried out [9–11]. Zentar et al. [12] conducted experimental investigations into the tensile strength and unconfined compressive strength of solidified marine sediments using siliceous-aluminous fly ash and cement. Through laboratory-unconfined compression tests, split tensile tests, bender element tests and isotropic compression tests conducted by Xiao et al. [13], a semiempirical relationship between compressive strength and curing time for fly ash-blended cement-stabilized marine clay has been obtained. A series of laboratory experiments, including isotropic compression, triaxial drained shearing, unconfined compression and bender element testing were carried out by Cheng et al. [14], in which the primary yielding and yield locus of fly ash cement-stabilized marine clay were investigated.

Polypropylene fiber is characterized by light weight, high tensile strength and low energy consumption. The strength and deformation resistance of cement-stabilized soils can be improved by incorporating appropriate fibers [15–17]. Many studies have been carried out on the properties of polypropylene fiber-reinforced cement-stabilized soils. Correia et al. [18] conducted experiments and concluded the compressive and tensile strength characteristics of polypropylene fiber-reinforced blast furnace slag solidified Portuguese soft soils. Through laboratory-split tensile tests conducted by Xiao et al. [19]—as well as stochastic finite element theory—a prediction model for tensile strength of polypropylene / polyvinyl alcohol fiber-modified cement-stabilized clay was established considering the fiber length and content. Ding et al. [20] studied the effect of freeze-thaw cycles on the mechanical properties of polypropylene fiber-reinforced cement-stabilized clay; the relationships among sample size, residual stress ratio, tangent modulus, cement content, fiber content and number of freeze-thaw cycles were established. The tensile strength of cement-stabilized marine clay reinforced by short waste fibers was investigated by Li et al. [21]. Through experimental and numerical methods, a numerical simulation of a single fiber pullout from a matrix was established by using a cohesive contact model.

Compared with the prevalence of studies on unconfined compression strength, there have been fewer studies on the compressive characteristics of fiber-reinforced fly ash-cemented clay. This paper presents a study on the effect of polypropylene fiber content, composite cement content and curing time on the unconfined compression strength of polypropylene fiber-composite cement-reinforced marine clay by using experimental tests, in order to provide reference for the future application of polypropylene fiber in soft clay foundations.

2. Materials and Experimental Methodology

2.1. Materials

The studied materials mainly included marine clay, fly ash cement and polypropylene fibers. The clay used in the test was Singapore marine clay excavated from a subway station with characteristics of grayish brown, saturated, obvious rheological properties, high compressibility and low bearing capacity. The main physical characteristics of the marine clay are presented in Table 1. The marine clay was put into a PVC plastic shading bucket and transported to the laboratory for backup. The water content of marine clay was approximately 88%. The liquid limit and plastic limit of marine clay were, respectively, 73% and 32%.

Table 1. Physical characteristics of marine clay.

Water Content (%)	Specific Gravity, Gs (Mg/m^3)	Liquid Limit (%)	Plastic Limit (%)	Plasticity Index (%)
88	2.7	73	32	41

The test cement is CEM II / B-V commercial composite cement and its physical and mechanical indexes are shown in Table 2. The mass ratio of fly ash to ordinary cement in CEM II / B-V commercial composite cement is 0.345:0.655; its main chemical components are SiO_2, Al_2O_3, Fe_2O_3, MgO and CaO.

Table 2. Indexes of mechanical properties of fly ash cement.

Initial Setting Time (h)	Final Setting Time (h)	Stability	Loss on Ignition (%)	Compressive Strength (MPa)		Flexural Strength (MPa)	
				3 d	28 d	3 d	28 d
2:30	4:00	Qualified	1.3	24	43.0	4.8	7.2

The polypropylene fibers used in the test were chopped 6-mm commercial fibers. The main physical and mechanical characteristics of polypropylene fibers are shown in Table 3.

Table 3. Physical and mechanical characteristics of polypropylene fibers.

Type	Density (g/cm^3)	Length (mm)	Tensile Strength (MPa)	Acid and Alkali Resistance	Dispersion
Monofilament	0.91	6	>360	High performance	Good

2.2. Experiment Scheme

The definition of compound composite content, polypropylene fiber content and water content are shown in Equations (1)–(3), respectively. The fly ash cement content was adopted based on applications in practice, such as the improvement of an ending shaft to supporting a tunnel boring machine. The fiber content was determined as the common values used in foundation pit project and subgrade engineering.

$$C_c = M_c/M_s \times 100\% \tag{1}$$

$$C_f = M_f/M_s \times 100\% \tag{2}$$

$$C_w = M_w/(M_s + M_c) \times 100\% \tag{3}$$

where C_c is fly ash cement content, C_f is polypropylene fiber content, C_w is water content, M_c is the quality of fly ash cement, M_s is the quality of dry soil in the marine clay, M_f is the quality of polypropylene fiber, M_w is the quality of water.

An orthogonal design method was used to study the influence of factors such as fiber content, composite cement content and curing time on the unconfined compressive strength of polypropylene fiber-reinforced fly ash-cemented marine clay. The test scheme is shown in Table 4. Considering test error factors such as sample heterogeneity, 5 samples were conducted in each group.

Table 4. The test scheme of the unconfined compressive strength.

Polypropylene Fiber Content (%)	Fly Ash Cement Content (%)	Curing Time (d)	Water Content (%)
	20, 50, 100	7	
0	20, 50, 100	14	
	20, 50, 100	28	
	20, 50, 100	7	
0.5	20, 50, 100	14	100
	20, 50, 100	28	
	20, 50, 100	7	
1	20, 50, 100	14	
	20, 50, 100	28	

2.3. Sample Preparation and Testing

The obvious plant roots, shells and other debris were removed from the marine clay at the site. Moreover, the sand particles are sieved with a 1-mm sieve. Samples were prepared in accordance with a procedure described by Chin et al. [22]. The reshaping of marine clay and the preparation method of triaxial samples were consistent with references [6,14,23]. According to the content of polypropylene fiber, composite cement and water in the test scheme, the required quality of marine clay, composite cement, fiber and water were weighed with a high-precision electronic scale. First, appropriate amount of water was added to the marine clay and stirred well. Secondly, the weighed composite cement was added and stirred well. Finally, the weighted polypropylene fibers were added and stirred well. The stirred fiber-reinforced fly ash cemented clay was put into a sealed plastic bag for later use.

The PVC plastic mold with an inner diameter of 50 mm and a height of 100 mm was fastened and cleaned. One end of the mold with a plastic film was sealed and then a release agent evenly inside the mold was applied. The mixture of fiber-composite cement-reinforced clay was stirred evenly into the mold for 5 times. We ensured the mixture previously squeezed in was shaken evenly before the next squeeze. After the top of the sample that has been shaken uniformly was scraped with a spatula, the two ends of the mold were completely wrapped with labeled water-permeable filter paper and placed horizontally in the conservational water tank for indoor conservation to ensure that the specimen was not affected by any external force during the conservational process.

The cylindrical samples were taken out from the conservational water tank and wiped off the surface water when the curing time reached 7 d, 14 d and 28 d, respectively. The experiment of unconfined compressive strength was conducted using a triaxial test device with a load at a constant rate of 1 mm/min after the samples were scraped well. The relevant data during the test were collected. An unconfined compression test was performed by following the procedures prescribed in ISO/TS 17892 (2004) [24].

3. Results and Discussion

3.1. Stress-Strain Behavior

Figures 1–3 present the curves of axial strain $\varepsilon - \delta$ with 0%, 0.5% and 1% polypropylene fiber content of the fiber-reinforced fly ash-cemented marine clay, respectively, depending on different mix ratio and curing time. All stress-strain curves show the peak strength of fiber-reinforced fly ash-cemented marine clay was significantly higher than that without polypropylene fiber, except the condition of 20% compound fly ash cement content. Compared with fly ash cement-reinforced marine clay, the residual strength deformation stage of the stress-strain curve of fiber-reinforced composite-cemented clay was more significant and the retention time of residual strength was much longer after reached the peak strength. Taking the 50% fly ash cement content and 0.5% fiber content as an example, the residual strength of the fiber-reinforced fly ash-cemented marine clay increased by 133.3%, 145.6% and 224.5% when the curing age was 7 d, 14 d and 28 d, respectively, which shows that adding certain quality polypropylene fiber can effectively improve the residual strength of fly ash cement-reinforced marine clay. Moreover, the trend for the changes of curves of axial strain $\varepsilon - \delta$—as well as strength and stiffness without fiber with fly ash cement content and curing time—is similar to the observation reported in a previous study [6,9,14]. The researchers analyzed the structure with respect to the unconfined compression strength of fly ash cement-reinforced clay.

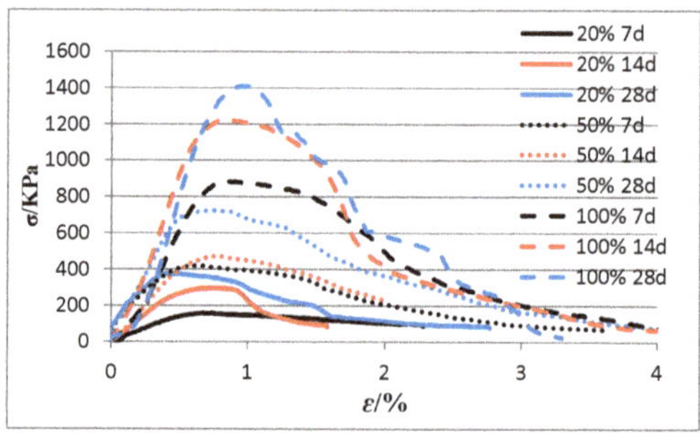

Figure 1. Curves of axial strain $\varepsilon - \delta$ without fiber depending on fly ash cement content and curing time.

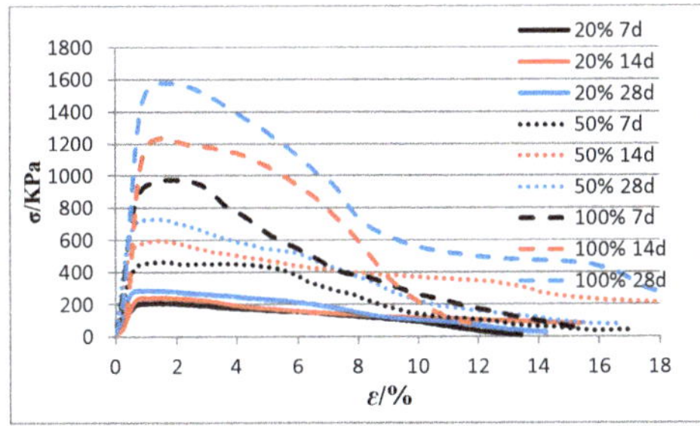

Figure 2. Curves of axial strain $\varepsilon - \delta$ with 0.5% fiber content depending on fly ash cement content and curing time.

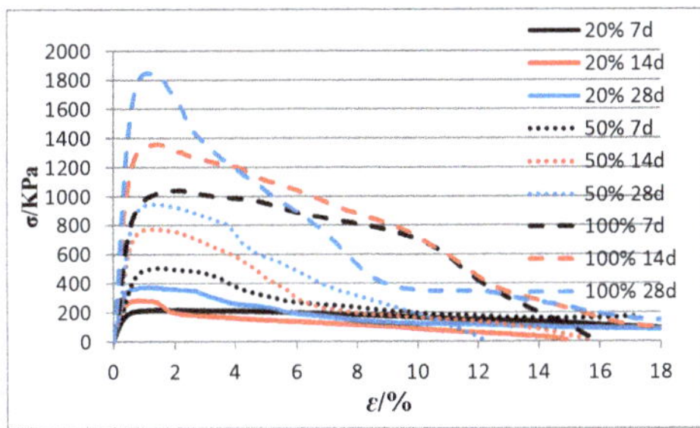

Figure 3. Curves of axial strain $\varepsilon - \delta$ with 1% fiber content depending on fly ash cement content and curing time.

The axial strain corresponding to the failure time of fiber-reinforced composite-cemented marine clay is shown in Table 5. Taking the 50% fly ash cement content and 0.5% fiber content as an example, the axial strain at the failure time of the fiber-reinforced fly ash-cemented marine clay increased by 253.5%, 197.3% and 172.9% when the curing age is 7 d, 14 d and 28 d, respectively. The axial strain at the time of failure of fiber-reinforced composite-cemented marine clay was much larger than that without polypropylene fiber under the same conditions, which indicates that the incorporation of polypropylene fiber can effectively improve the brittleness and toughness of the composite cement-reinforced marine clay. The main reason is that the composite cement-reinforced marine clay particles attached to the fiber surface can increase its cohesion and friction. The tensile stress between the fiber and the reinforced clay may continue for a long time though the specimen is already damaged, which means that the fiber can effectively reduce the deformation and improve the toughness and residual strength of the fly ash cement-reinforced marine clay.

Table 5. The axial strain at failure time of the fiber fly ash cement-reinforced marine clay.

Fly Ash Cement Content (%)	Curing Time (d)	Polypropylene Fiber Content (%)		
		0	0.5	1
	7	0.668	1.474	1.604
20	14	0.716	1.566	0.797
	28	0.4	0.778	0.908
	7	0.63	1.597	1.427
50	14	0.744	1.468	1.24
	28	0.727	1.257	1.397
	7	0.865	1.855	2.17
100	14	0.807	1.579	1.466
	28	0.952	1.483	1.083

3.2. Effect of Fiber Content on Unconfined Compressive Strength

The relationship between the unconfined compressive strength and the fiber content of each group of samples depending on fly ash cement content and curing time is shown in Figure 4. The figure indicates that the unconfined compressive strength of polypropylene fiber-reinforced composite-cemented marine clay increases at first and then decreases with the increase of fiber content when the composite cement content is 20%, however, it keeps increase with the increase of fiber content when the composite cement content is 50% and 100%. Taking the 50% fly ash cement content and 0.5% fiber content as an example, the unconfined compressive strength of the fiber-reinforced fly ash-cemented marine clay at curing time 7 d, 14 d and 28 d is 1.11 times, 1.27 times and 1.06 times, respectively, that of without polypropylene fiber. When the fiber content is 1% with the 50% fly ash cement content, the unconfined compressive strength of the fiber-reinforced fly ash-cemented marine clay is 1.20 times, 1.63 times and 1.31 times that of without polypropylene fiber, which indicates that with low fly ash cement content, the structure of composite cement-reinforced marine clay is not fully formed; the modification effect of polypropylene fiber on composite cement-reinforced clay is not obvious. However, when the content of fly ash cement is higher, adding appropriate amount of fiber can effectively increase its cohesion of the fly ash cement-reinforced marine clay, which can make the reinforcing effect of polypropylene fiber obvious. The mechanical performance of fiber-composite fly ash cement-reinforced clay with 0.5% fiber content is basically consistent with previous studies, however, it shows different characteristics with 1% fiber content [16,17,22].

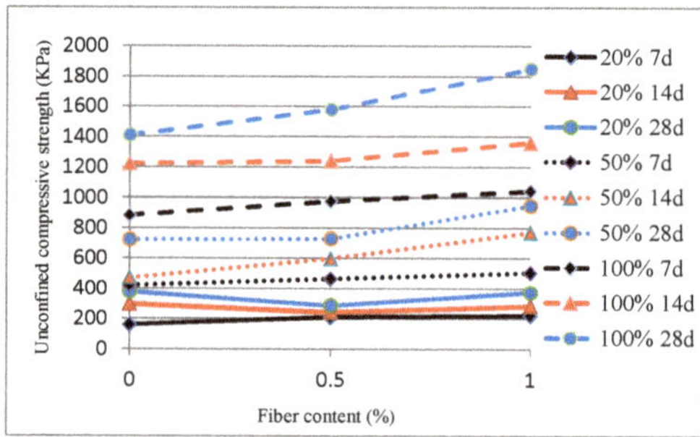

Figure 4. Relationship between unconfined compressive strength and fiber content depending on fly ash cement content and curing time.

3.3. Effect of Fly Ash Cement Content on Unconfined Compressive Strength

In order to study the effect of fly ash cement content on the unconfined compressive strength, the relationship depending on fiber content and curing time was studied, as shown in Figure 5. The figure indicates that the unconfined compressive strength of polypropylene fiber-reinforced fly ash-cemented marine clay increased with the increase of fly ash cement content. The increase trend of 14 d to 28 d was more significant than that from 7 d to 14 d. Taking the 0.5% fiber content as an example, the unconfined compressive strength of 50% fly ash cement content of the fiber-reinforced fly ash-cemented marine clay at curing time 7 d, 14 d and 28 d was 2.23 times, 2.47 times and 2.52 times that of 20% compound cement content respectively. As a comparison, the unconfined compressive strength of 100% fly ash cement content of the fiber-reinforced fly ash-cemented marine clay at curing time 7 d, 14 d and 28 d was 4.67 times, 5.13 times and 5.48 times, respectively, that of 20% compound cement content. It was shown that the fly ash cement content had a significant effect on the unconfined compressive strength of fiber-reinforced composite-cemented marine clay: the unconfined compressive strength increased significantly with the increase of curing time and the polypropylene fiber content.

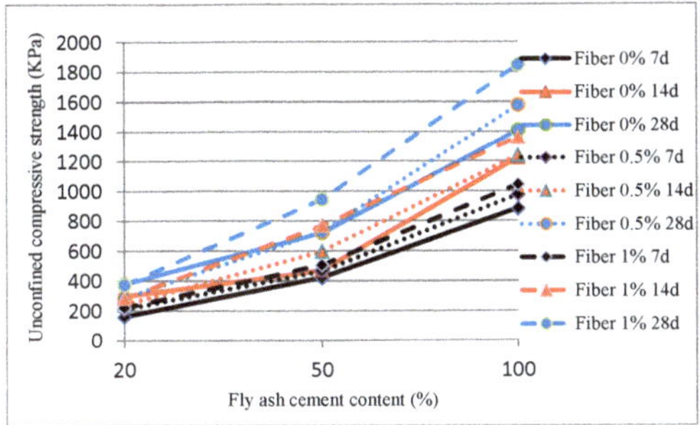

Figure 5. Relationship between unconfined compressive strength and fly ash cement content depending on fiber content and curing time.

3.4. Effect of Curing Time on Unconfined Compressive Strength

The relationship between the unconfined compressive strength and the curing time depending on different fiber content and fly ash cement content is shown in Figure 6. The figure indicates that the unconfined compressive strength of polypropylene fiber-reinforced fly ash-cemented marine clay increased with the increase of curing time. Moreover, the growth rate increased more obviously when the curing time increased. Taking the 50% fly ash cement content and 0.5% fiber content as an example, the unconfined compressive strength of the fiber-reinforced fly ash-cemented marine clay at curing time 14 d and 28 d was 1.29 times and 1.57 times, respectively, than that of 7 d curing time. The relationship between the unconfined compressive strength and the curing time at 7 d, 14 d and 28 d can be expressed by Equations (4)–(6), respectively.

$$f_{ts} = 176.4e^{0.16t} \tag{4}$$

$$f_{ts} = 373.9e^{0.23t} \tag{5}$$

$$f_{ts} = 373.9e^{0.23t} \tag{6}$$

where f_{ts} is unconfined compressive strength, t is curing time.

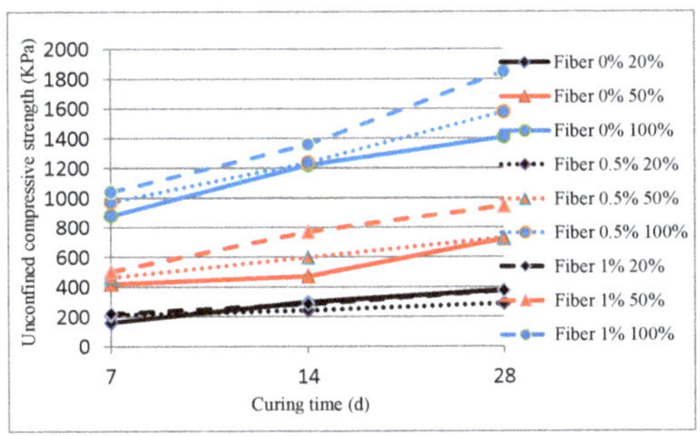

Figure 6. Relationship between unconfined compressive strength and curing time, depending on fiber content and fly ash cement content.

3.5. Failure Modes

Taking the 7 d curing time as an example, the failure photos of fiber-composite cement-reinforced marine clay with 0.5% polypropylene fiber content are shown in Figure 7.

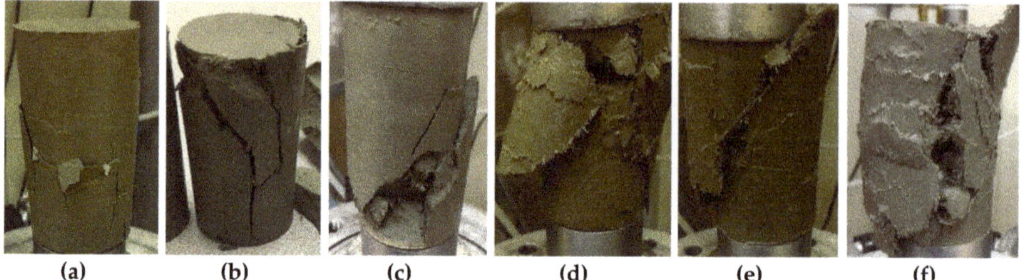

(a)   (b)   (c)   (d)   (e)   (f)

Figure 7. Failure photos of the specimens with 0.5% fiber content and 7 d curing time. (**a**) 20% fly ash cement content; (**b**) 50% fly ash cement content; (**c**) 100% fly ash cement content; (**d**) 20% fly ash cement content with fiber; (**e**) 50% fly ash cement content with fiber; (**f**) 100% fly ash cement content with fiber.

As can be seen from Figure 7, the unconfined compressive strength of the fly ash cement-reinforced marine clay increased with the increase of fly ash cement content; the failure of the specimen gradually showed the characteristics of brittle failure. When 20% fly ash cement was added, there were no obvious through-cracks in the fly ash cement-reinforced clay. However, samples with 50% and 100% fly ash cement content showed obvious through-cracks in a short period of time; specimens with 100% fly ash cement content showed surface shedding during compression.

Compared with fly ash-cemented marine clay, fiber-reinforced fly ash-cemented marine clay had a longer crack generation time and a slower crack propagation speed under the same conditions. The specimen of fiber-reinforced composite-cemented marine clay showed good plastic deformation after compaction and failure without phenomenon of surface shedding. The tensile stress between the fiber and the fly ash cement-reinforced clay could continue for a long time under the continuous compressive stress though the specimen was already broken. It was affected by the mechanical behavior of the interface between the fiber and the soil particle matrix microscopically. The results were similar to those in reference [16]. The specimen cracked, instead of broke, macroscopically.

4. Conclusions

The following conclusions can be drawn according to the tests results:

The peak strength of fiber-reinforced fly ash-cemented marine clay is significantly higher than that without polypropylene fiber except on the condition of 20% fly ash cement content. The axial strain at the time of failure of fiber-reinforced composite-cemented marine clay is much larger than that without polypropylene fiber under the same conditions, which indicates that the incorporation of polypropylene fiber can effectively improve the peak strength and toughness of the composite cement-reinforced marine clay.

The unconfined compressive strength of fiber-reinforced fly ash-cemented marine clay shows different trends with the increase of fly ash cement content: it increases at first, and then decreases with 20% fly ash cement content, while it appears an increasing trend with 50% and 100% fly ash cement content. The unconfined compressive strength of fiber-reinforced composite-cemented marine clay increases with the increase of fly ash cement content and curing time. The increase trend of the unconfined compressive strength from 14 d to 28 d is significantly larger than that from 7 d to 14 d. The growth rate of the unconfined compressive strength increases with the increase of the curing time and the unconfined compressive strength shows a growth trend of power function.

Composite cemented marine clay forms through-cracks in a short time and shows a brittle failure mode during compression, while fiber-reinforced fly ash-cemented marine clay shows a plastic failure mode which appears that it can continue for a long time after the formation of through cracks. On the microscopic level, it shows continuous tensile stress between polypropylene fiber and composite cement-reinforced marine clay; on the macro level, it became cracked and continuous.

Author Contributions: Q.C. and J.Z. conceived and designed the study; Q.C. and N.Z. performed the experiments and wrote the paper; Y.G. and S.P. reviewed and edited the manuscript. All authors have read and agreed to the published version of the manuscript.

Funding: This work is funded by the Research Fund of The State Key Laboratory of Coal Resources and Safe Mining, CUMT (SKLCRSM19KF012); Qing Lan Project (2018); Ministry of Housing and Urban-Rural Development Technology Plan (2018-K7-004); Science and Technology Project of Construction System in Jiangsu Province (2018ZD021, 2019ZD083); and the Doctoral Special Fund of Jiangsu Vocational Institute of Architectural Technology (JYBZX18-02).

Acknowledgments: The authors gratefully acknowledge the support provided by National University of Singapore and the Jiangsu Overseas Research & Training Program for University Prominent Young & Middle-aged Teacher and Presidents.

Conflicts of Interest: All the authors declare that there are no conflicts of interest regarding the publication of this paper.

References

1. Liu, Y.; Jiang, Y.; Xiao, H.; Lee, F.H. Determination of representative strength of deep cement-mixed clay from core strength data. *Géotechnique* **2017**, *67*, 350–364. [CrossRef]
2. Consoli, N.C.; Da Fonseca, A.V.; Silva, S.; Cruz, R.; Fonini, A. Parameters controlling stiffness and strength of artificially cemented soils. *Géotechnique* **2012**, *62*, 177–183. [CrossRef]
3. Kamruzzaman, A.H.; Chew, S.H.; Lee, F.H. Structuration and Destructuration Behavior of Cement-Treated Singapore Marine Clay. *J. Geotech. Geoenviron. Eng.* **2009**, *135*, 573–589. [CrossRef]
4. Lee, F.H.; Lee, Y.; Chew, S.-H.; Yong, K.-Y. Strength and Modulus of Marine Clay-Cement Mixes. *J. Geotech. Geoenviron. Eng.* **2005**, *131*, 178–186. [CrossRef]
5. Liu, Y.; Chen, E.J.; Quek, S.-T.; Yi, J.-T.; Lee, F.H. Effect of spatial variation of strength and modulus on the lateral compression response of cement-admixed clay slab. *Géotechnique* **2015**, *65*, 851–865. [CrossRef]
6. Xiao, H. Evaluating the Stiffness of Chemically Stabilized Marine Clay. *Mar. Georesources Geotechnol.* **2016**, *35*, 698–709. [CrossRef]
7. Xiao, H.; Lee, F.H.; Liu, Y. Bounding Surface Cam-Clay Model with Cohesion for Cement-Admixed Clay. *Int. J. Géoméch.* **2017**, *17*, 04016026. [CrossRef]
8. Liu, Y.; He, L.Q.; Jiang, Y.J.; Sun, M.M.; Chen, E.J.; Lee, F.-H. Effect of in situ water content variation on the spatial variation of strength of deep cement-mixed clay. *Géotechnique* **2019**, *69*, 391–405. [CrossRef]
9. Xiao, H.W.; Wang, W.; Goh, S.H. Effectiveness study for fly ash cement improved marine clay. *Constr. Build. Mater.* **2017**, *157*, 1053–1064. [CrossRef]
10. Show, K.-Y.; Tay, J.-H.; Goh, A. Reuse of Incinerator Fly Ash in Soft Soil Stabilization. *J. Mater. Civ. Eng.* **2003**, *15*, 335–343. [CrossRef]
11. Zhou, N.; Ouyang, S.; Cheng, Q.; Ju, F. Experimental Study on Mechanical Behavior of a New Backfilling Material: Cement-Treated Marine Clay. *Adv. Mater. Sci. Eng.* **2019**, *2019*, 1–8. [CrossRef]
12. Zentar, R.; Wang, D.; Abriak, N.-E.; Benzerzour, M.; Chen, W. Utilization of siliceous–aluminous fly ash and cement for solidification of marine sediments. *Constr. Build. Mater.* **2012**, *35*, 856–863. [CrossRef]
13. Xiao, H.; Shen, W.; Lee, F.H. Engineering Properties of Marine Clay Admixed with Portland Cement and Blended Cement with Siliceous Fly Ash. *J. Mater. Civ. Eng.* **2017**, *29*, 04017177. [CrossRef]
14. Cheng, Q.; Xiao, H.; Liu, Y.; Wang, W.; Jia, L. Primary yielding locus of cement-stabilized marine clay and its applications. *Mar. Georesources Geotechnol.* **2018**, *37*, 488–505. [CrossRef]
15. Anggraini, V.; Huat, B.B.K.; Asadi, A.; Nahazanan, H. Effect of Coir Fibers on the Tensile and Flexural Strength of Soft Marine Clay. *J. Nat. Fibers* **2014**, *12*, 185–200. [CrossRef]
16. Tang, C.-S.; Shi, B.; Gao, W.; Chen, F.; Cai, Y. Strength and mechanical behavior of short polypropylene fiber reinforced and cement stabilized clayey soil. *Geotext. Geomembr.* **2007**, *25*, 194–202. [CrossRef]
17. Kaniraj, S.R.; Havanagi, V.G. Behavior of Cement-Stabilized Fiber-Reinforced Fly Ash-Soil Mixtures. *J. Geotech. Geoenviron. Eng. ASCE.* **2001**, *127*, 574–584. [CrossRef]
18. Correia, A.A.S.; Paulo, J.V.O.; Custódio, D.G. Effect of polypropylene fibers on the compressive and tensile strength of a soft soil, artificially stabilised with binders. *Geotext. Geomemb.* **2015**, *43*, 97–106. [CrossRef]
19. Xiao, H.; Liu, Y. A prediction model for the tensile strength of cement-admixed clay with randomly orientated fibers. *Eur. J. Environ. Civ. Eng.* **2016**, *22*, 1131–1145. [CrossRef]

20. Ding, M.; Zhang, F.; Ling, X.; Lin, B. Effects of freeze-thaw cycles on mechanical properties of polypropylene Fiber and cement stabilized clay. *Cold Reg. Sci. Technol.* **2018**, *154*, 155–165. [CrossRef]

21. Li, Q.; Chen, J.; Hu, H. The tensile and swelling behavior of cement-stabilized marine clay reinforced with short waste fibers. *Mar. Georesources Geotechnol.* **2019**, *37*, 1236–1246. [CrossRef]

22. Chin, K.G.; Lee, F.H.; Dasari, G.R. Effects of curing stress on mechanical properties of cement-treated soft marine clay. In Proceedings of the International Symposium on Engineering Practice and Performance of Soft Deposits, Toyonaka, Japan, 2–4 June 2004; pp. 217–222.

23. Cheng, Q.; Yao, K.; Liu, Y. Stress-Dependent Behavior of Marine Clay Admixed with Fly-Ash-Blended Cement. *Int. J. Pavement. Res. Technol.* **2018**, *11*, 611–616. [CrossRef]

24. *Geotechnical Investigation and Testing—Laboratory Testing of Soil— Part 7*; ISO/TS 17892; International Organization for Standardization: Geneva, Switzerland, 2004.

Article

Influence of Combined Action of Steel Fiber and MgO on Chloride Diffusion Resistance of Concrete

Feifei Jiang [1,2,*], Min Deng [1], Liwu Mo [1] and Wenqing Wu [3]

[1] State Key Laboratory of Materials-Oriented Chemical Engineering, College of Materials Science and Engineering, Nanjing Tech University, Nanjing 211800, China; dengmin@njtech.edu.cn (M.D.); andymoliwu@njtech.edu.cn (L.M.)

[2] College of Naval Architecture Civil Engineering, Jiangsu University of Science and Technology, Zhangjiagang Campus, Suzhou 215600, China

[3] School of Transportation, Southeast University, Nanjing 210089, China; wuwenqing@seu.edu.cn

* Correspondence: 999620140019@just.edu.cn

Received: 24 March 2020; Accepted: 22 April 2020; Published: 24 April 2020

Abstract: To improve the chloride diffusion resistance and durability of concrete, a new kind of steel fiber reinforced MgO concrete (SFRMC) was made by adding steel fiber and MgO to concrete simultaneously. With steel fiber for load bearing and expansion limiting, MgO as the expander, SFRMC has both the advantages of fiber reinforced concrete and expansion concrete. The influence of steel fiber and MgO on the strength and chloride diffusion resistance of concrete was evaluated by splitting tensile test and chloride diffusion test. Mercury intrusion porosimeter (MIP) and scanning electron microscopy (SEM) were used to study the microstructure of SFRMC. The results showed that the combined action of steel fiber and MgO reduced the porosity of concrete and the chloride diffusion coefficient (CDC), which could not be achieved by steel fiber and MgO separately. In the free state, the expansion energy produced by the hydration of MgO made the concrete expand outwards. However, under the constraint of steel fiber, the expansion energy was used to tension the fiber, resulting in self-stress. In this way, compared to reference concrete RC, the tensile strength of SFRMC-1, SFRMC-2, and SFRMC-3 increased by 3.1%, 61.3%, and 64.5%, CDC decreased by 8.8%, 36.7%, and 33.1%, and the porosity decreased by 6.2%, 18.4%, and 20.6%, respectively. In addition, the SEM observations demonstrated that the interfacial transition zone (ITZ) between fiber and matrix was denser in SFRMC, which contributed to reduce the diffusion of chloride ions in the concrete.

Keywords: steel fiber; MgO expansive agent; split tensile strength; chloride diffusion resistance; porosity; interfacial transition zone

---

1. Introduction

Steel fiber reinforced concrete (SFRC) is a kind of high-performance concrete which distributes steel fiber uniformly in the matrix. Due to the restraint of steel fibers, the development of cracks is restrained, which makes SFRC have higher crack resistance and good toughness [1]. Owing to these excellent properties, SFRC has been widely used in port structures and protection structures, especially in those structures with high requirements for diffusion resistance and crack resistance [2–4]. However, these structures mentioned above are often in direct contact with chlorine ions. When chloride ions diffuse into the concrete, they will cause corrosion of the steel bar, leading to cracking and affecting the safety of the structure [5,6]. Therefore, it is of great significance to reduce the chloride diffusion coefficient (CDC) of SFRC.

In recent years; many researchers have studied the chloride diffusion resistance of fiber reinforced concrete. Vahid compared the effects of different kinds of fibers on the resistance of concrete to chloride tolerance. He found that polypropylene fiber reduced CDC; while adding steel fibers significantly increased CDC [7]. Guo added 0.15%, 0.30%, 0.45%, 0.60% of basalt fiber into concrete and tested the pore structure and the chloride diffusion resistance. The results showed that by adding basalt fiber, the chloride diffusion resistance was improved, the minor harmful pores (20–100 nm) were increased, and the serious harmful pores (>200 nm) were significantly reduced [8]. Yan explored the effects of different amounts of basalt fiber (BF; 0.05, 0.1, 0.15, and 0.2 vol%) on the chloride ion diffusion. He found that the inclusion of 0.05% BF accelerated the diffusion of chloride ions from the coral aggregate; while a dosage of BF above 0.1% could suppress the diffusion of chloride ions [9]. Mahyuddin explored the effects of different amounts of coconut fiber (0.6%, 1.2%, 1.8%, and 2.4%) on mechanical property and chloride diffusion. He found that by adding coconut fiber, compressive and flexural strengths were increased to 13% and 9%, respectively. However, in terms of durability, the chloride diffusion resistance was reduced [10]

Previous studies have confirmed that fiber can improve the mechanical properties of concrete [11–16], but there are different opinions on the chloride diffusion resistance. Some researchers think that fiber can enhance the chloride diffusion resistance of concrete, but the other researchers hold the opposite opinion. The main reason for different opinions is that fiber has both positive and negative effects in concrete. On the one hand, fiber reduces the shrinkage crack of concrete and improves the compactness of concrete. On the other hand, the fiber increases the number of ITZ between the fiber and the matrix, and the gap at the interface provides a channel for chloride diffusion. However, researchers have agreed that ITZ between fiber and matrix is the weakness of concrete, and the strength of ITZ has a great influence on the performance of concrete. To reduce the defect of ITZ, steel fiber and MgO were used simultaneously in this paper. The expansion of MgO was restrained by steel fiber, the void was filled, the porosity was reduced, and the strength of ITZ was improved.

From the view of reducing shrinkage cracks, the use of expansion agent is to use the expansion produced by hydration to compensate the shrinkage of concrete [17]. According to the differences in mineral composition of the expansive agent, the expansive agents currently sold in the market can be divided into five types: sulfide-aluminate expansive agent, lime expansive agent, iron powder expansive agent, MgO expansive agent, and compound expansive agent [18]. Among the above, MgO is one of the most excellent expansive agents. It has been widely used because of its stable hydration products, easy regulation of expansion performance, and no shrinkage in the later stage [19,20].

By studying the advantages and disadvantages of fiber and MgO, we found that steel fiber and MgO can be used together to give full play to the advantages of both materials. Different from SFRC, steel fiber not only plays a role in bridge connection, but also plays a role in limiting expansion in SFRMC. The three-dimensional distribution of steel fiber seriously restricts the expansion of MgO, changes the expansion mode of MgO, and makes the expansion from outward extension to inward extrusion, which greatly improves the compactness of SFRMC. Our previous research [21] has confirmed that SFRMC has excellent mechanical properties, but its durability has not been studied. In this paper, split tensile test, chloride diffusion test, mercury intrusion porosimeter MIP test, and scanning electron microscopy SEM test were conducted to study the combined effect of steel fiber and MgO on the tensile strength and chloride diffusion resistance of SFRMC. Through the analysis of pore structure, ITZ structure, the mechanism of SFRMC performance enhancement was studied. These data provided theoretical support for further application of SFRMC and opened up new application fields for MgO concrete and fiber reinforced concrete.

2. Materials and Methods

2.1. Materials

In this paper, the cement is Class 52.5 Ordinary Portland Cement (Shan Aluminum Cement Co., Ltd., Shandong, China). Fly ash was produced by Shenhua Huashou Power Co., Ltd in Shanghai, China. MgO was produced by Wuhan Sanyuan Special Building Materials co. LTD in Wuhan, China.

The activity of MgO was 115 s and the specific surface area was 45.7 m^2/g. Figure 1 shows the mineral composition of MgO. Table 1 summarizes the chemical composition of cement, Fly ash, and MgO.

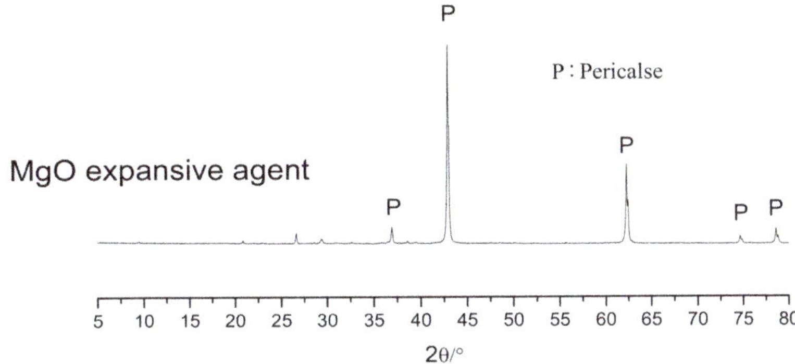

Figure 1. X-ray diffraction pattern of MgO.

Table 1. Chemical composition of cement.

Type	Chemical Composition /wt%									
	CaO	MgO	Al$_2$O$_3$	SiO$_2$	Fe$_2$O$_3$	SO$_3$	K$_2$O	Na$_2$O	Loss	Total
Cement	60.51	2.18	6.34	22.02	3.05	1.86	0.47	0.23	1.96	98.62
Fly ash	5.01	1.03	34.18	48.91	5.22	1.20	0.89	0.62	1.50	98.56
MgO	3.19	85.44	0.73	4.45	0.42	0	0	0	4.49	98.72

Continuously graded gravel with a size of 5–20 mm was used as the coarse aggregates. River sand with a fineness modulus of 2.94 was used as the fine aggregates. Steel fiber was produced by Zibo Shuanglian Building Materials co. LTD in China. The tensile strength of steel fiber was 520 MPa. The diameter of steel fiber was 0.58 mm and the length was 40 mm. Figure 2 shows the specific morphology of steel fiber, which is a wave shape. Table 2 shows the mix proportion of the concrete used in this paper.

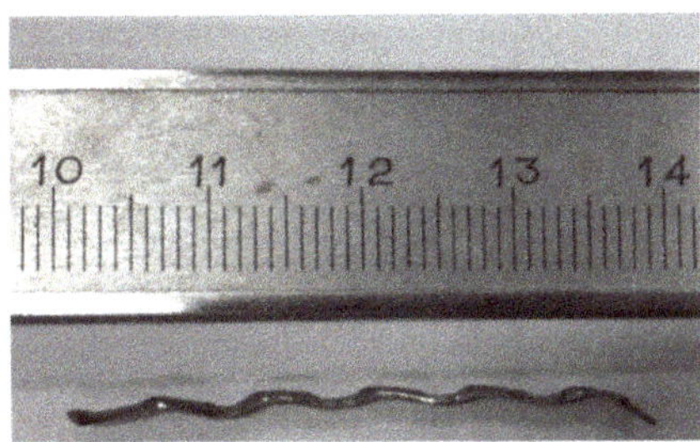

Figure 2. Photograph of steel fiber.

2.2. Experimental

The specific process and analysis method of this paper are shown in Figure 3. Cube blocks of 150*150*150 mm were poured to study the split tensile strength and cylinder specimens of ∅100×50 mm were poured to study the chloride diffusion resistance. After casting, the molds filled with fresh concrete were placed on the high-frequency vibration table to vibrate until no obvious bubbles escaped. After 24 h, the abrasives were removed and the concrete was transferred to the standard curing chamber (20 ± 2 °C , 95%RH).

Table 2. Mix proportion of concrete.

Specimen	Slump/ mm	Composition /kg·m⁻³							
		Cement	Fly Ash	Fine Aggregate	Coarse Aggregate	Water	Water Reducer	Steel Fiber	MgO
Ref (RC)	155	450	50	713	1025	160	6	0	0
8%MgO(MC)	132	450	50	713	1025	160	6	0	40
0.5%Fiber(SFRC)	141	450	50	713	1025	160	6	39	0
0.5%Fiber + 8%MgO (SFRMC-1)	128	450	50	713	1025	160	6	39	40
1%Fiber+8%MgO (SFRMC-2)	115	450	50	713	1025	160	6	78	40
1.5%Fiber + 8%MgO (SFRMC-3)	103	450	50	713	1025	160	6	117	40

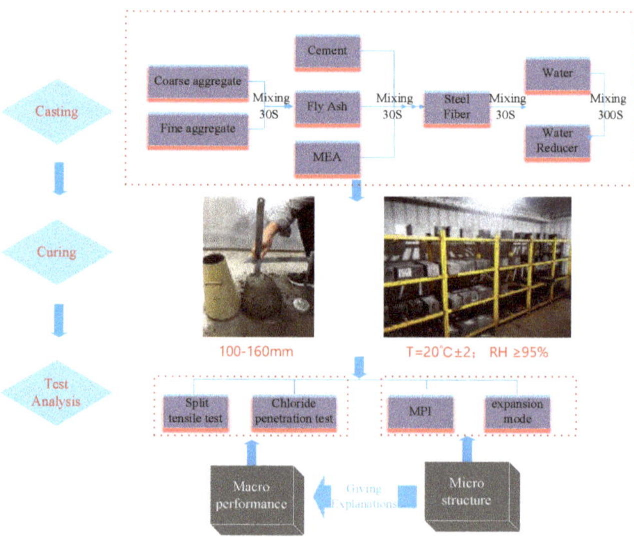

Figure 3. The specific process and analysis method.

2.2.1. Split Tensile Test

The split tensile strength test was conducted with reference to China Standard GB/T 50081-2016 [22]. The main equipment was the SYE-2000 pressure testing machine with a maximum load of 2000 kN. The size of the specimens was 150 mm × 150 mm × 150 mm, and the mean value of the three specimens was taken as the tensile strength of the specimens (accurate to 0.01 Mpa). The loading rate was 0.06 Mpa/s. The tensile strength was calculated according to Formula (1). The test diagram is shown in the Figure 4:

$$f_{ts} = 2F/\pi A = 0.637F/A, \tag{1}$$

where f_{ts} is the tensile strength of concrete, in MPa; F is the failure load of the test piece, in N; A is the bearing area of the test piece, in mm².

2.2.2. Chloride Diffusion Test

Normally, the main methods for measuring chloride diffusion resistance are slow method and fast method. Among them, the slow method is too time-consuming and seldom used in practice, while the fast method can measure the chloride diffusion resistance of materials in a short time. As the most typical of the rapid method, the electric flux method was first proposed by the Portland Cement Association of the United States in 1981 and has become the most widely used in the world. The principle of the method is to use the electric field to accelerate the ion transport. The ions penetrate through the specimen under the action of DC power. In this paper, the chloride diffusion resistance was evaluated by the flux transferred in accordance with the procedures described in ASTM C1202 [23]. The test device is shown in Figure 5.

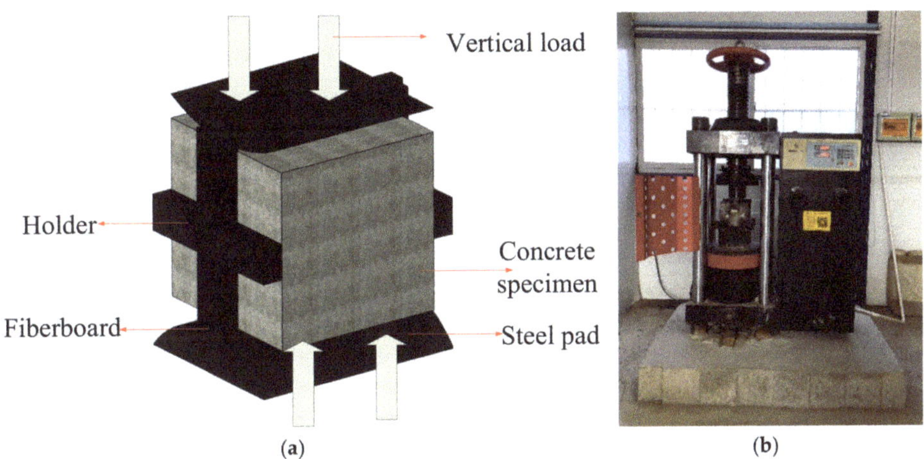

Figure 4. Setup for split tensile strength test: (**a**) Schematic diagram of test; (**b**) Photo of test.

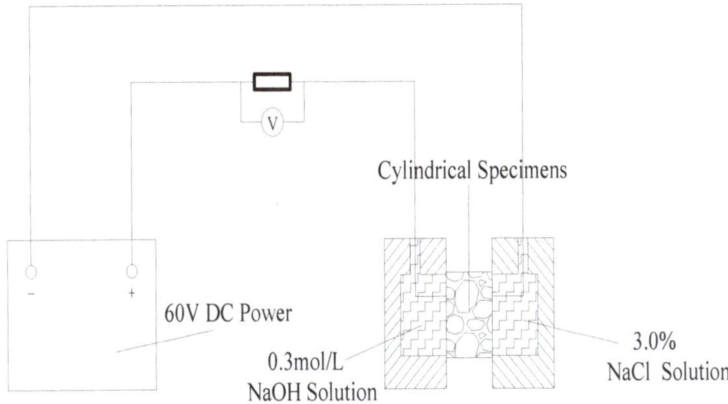

Figure 5. Setup for the chloride diffusion test.

Cylindrical specimens with a diameter of 100 mm and a height of 50 mm were used in the test. The solution was 0.3 *mol/l* NaOH and 3% NaCl. SX-DTL concrete chloride flux meter was used to record the electrification time and flux. The total electric flux of the concrete test block for 6 h was calculated according to Formula (2):

$$Q = 900 \times (I_0 + 2\,I_{30} + 2\,I_{60} + \cdots + 2\,I_t + \cdots + 2\,I_{300} + 2\,I_{330} + I_{360}), \tag{2}$$

where Q is total flux through test block for 6 h (C); I_0 is initial flux (A), to 0.001A; I_t is flux (A) at time t (min), to 0.001A.

According to the Nernst–Plank equation, the relationship between chloride diffusion coefficient (CDC) and Q was established (Formula (3)):

$$CDC = 2.57765 + 0.00492 \times Q. \tag{3}$$

2.2.3. MIP Test

The pore structure of concrete was analyzed by mercury intrusion porosimetry (MIP). When making the specimen, the concrete was knocked into several small test blocks with a length of about 2 mm to remove the coarse aggregate. The specimens were then soaked in anhydrous ethanol for 24 h to stop the hydration of cement and MgO. Then, the specimens were placed in a vacuum drying oven at 50 °C for drying for 12 h. Finally, the dried specimens were sealed in a plastic bag to prevent moisture from entering. During the test, the pore structure of two or three specimens was tested by MIP, and the effect of steel fibers and MgO on the pore structure was observed. By analyzing the experimental data of the pores, we could explain the variation law of tensile strength and compressive strength from the perspective of microstructure.

2.2.4. SEM Test

SEM test (JSM-6510lA, Japan) was used to study the interface performance between steel fiber and substrate. All specimens used for SEM test were standardly cured for 28 days.

3. Results

3.1. Failure Pattern of Concrete

At the beginning of the split tensile test, the specimen was in the elastic stage, the surface was intact, and the deformation increased proportionally with the increase of load. With the continuous increase of the load, the vertical cracks appeared in the middle of the specimen, and then gradually extended to the bottom and the top, accompanied by the sound of splitting. At the end of test, the specimen was completely damaged. There were no obvious signs before the failure of RC and MC. After the cracks appeared on the surface, they extended rapidly and destroyed the concrete completely, showing obvious brittle failure. Finally, the specimen broke into two independent parts (Figure 6a,b). When the steel fiber was added into concrete, owing to the steel fiber limited the crack extension, the crack developed slowly before the destruction of SFRC, showing obvious ductility. Some fine cracks appeared around the long crack, and the specimen remained as a complete whole after the test, with only a small amount of block concrete falling from the surface (Figure 6c).

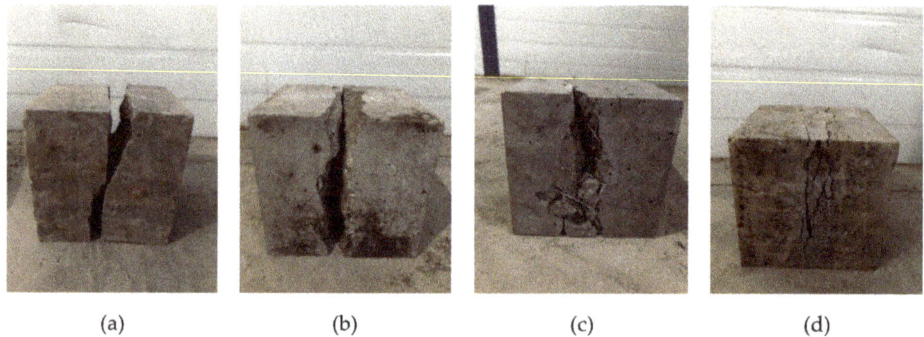

(a)   (b)   (c)   (d)

Figure 6. Failure pattern of concrete. (**a**) Reference concrete (RC); (**b**) MgO concrete (MC); (**c**) Steel fiber reinforced concrete (SFRC); (**d**) Steel fiber reinforced MgO concrete (SFRMC).

On the other hand, when the steel fiber and MgO were used simultaneously, the steel fiber restrained the expansion of MgO and generated self-stress, which improved the bite force between steel fiber and matrix, enhanced the strength of the interface between fiber and matrix, and significantly enhanced the crack resistance. Steel fiber made cracks became more evenly distributed. Instead of a single long and wide main crack in RC, the crack developed into many parallel fine cracks in SFRMC.

After the test, the SFRMC surface was smooth without concrete spalling, and the specimen split but did not separate (Figure 6d).

3.2. Combined Action of Steel Fiber and MgO on Split Tensile Strength

Figure 7 shows the split tensile strength of specimens with different mix proportions. Compared to RC, MgO had a small impact on the strength of MC in the early stage, but it had a greater impact on the strength in the later stage. At 28 days, the tensile strength of MC was 4.4% higher than that of RC, and the improvement was mainly due to the hydration of MgO. Compared to RC, the tensile strength of SFRC had been significantly improved. The strength had increased by 29.7% at 28 days, which was the result of the steel fiber restrained crack extension.

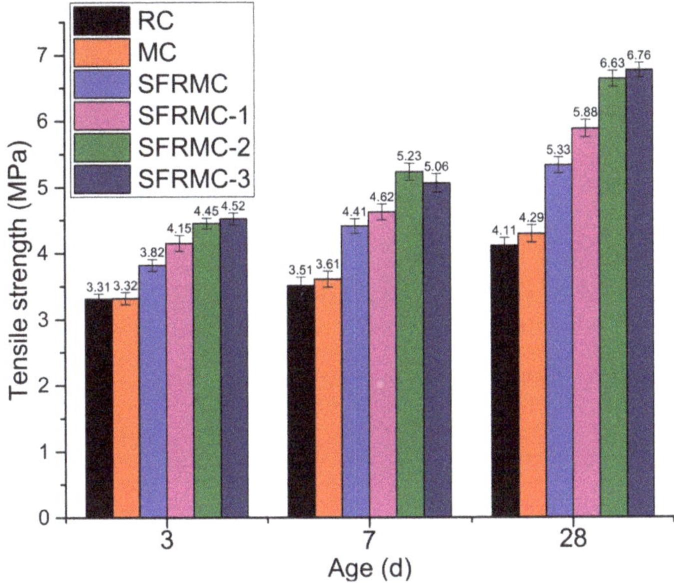

Figure 7. Split tensile strength of concrete.

On the other hand, when MgO and steel fiber were used at the same time, the tensile strength of SFRMC continued to increase. Compared to RC, the tensile strength of SFRMC-1, SFRMC-2, and SFRMC increased by 3.1%, 61.3%, and 64.5%, respectively. The tensile strength increased with the increase of steel fibers, and the growth rate was greater when the fiber was less than 1%, and it would no longer increase significantly when it exceeded 1%. Therefore, considering the cost of materials, we recommend using 8% MgO and 1% steel fiber in constructions.

As we can see from the split tensile test, the concrete strength had been significantly improved under the combined action of steel fiber and MgO. There are two main reasons for the increase. The first is that steel fiber limited the crack extension. When the cracks came out, the tensile stress was transferred from concrete to steel fiber, which restrained the further extension of cracks. The second reason is that the expansion of MgO was restrained by steel fiber, which resulted in self-stress. In the free state, the expansion energy produced by the hydration of MgO made the concrete expand outwards. However, under the constraint of steel fiber, the expansion energy was used to tension the fiber, resulting in self-stress. In this way, the tensile strength of concrete is obviously improved.

3.3. Combined Action of Steel Fiber and MgO on Chloride Diffusion Resistance

Table 3 shows the chloride diffusion coefficients (CDC) of concrete at 3, 7, 28, 60, and 180 days. As can be seen from Figure 8, with the increase in curing time, CDC of concrete with different mixing ratios gradually decreased, and the descending rate in the early stage (0–28 days) was relatively larger, while the rate in the later stage gradually decreased. The results showed that early curing of concrete had a significant effect on CDC. As the curing age increased, the cementitious materials continued to hydrate, the densification of concrete became higher, and the pores in concrete were gradually reduced, which was beneficial to reduce CDC and improve the durability. No matter how we changed the amount of fiber and MgO, the test results always obeyed this rule. Therefore, we suggest that in constructions, it is necessary to increase the time of early maintenance of concrete to improve the durability.

Table 3. Chloride diffusion coefficient of concrete specimens (10^{-9} cm^2/s).

Type	Curing Age				
	3 d	7 d	28 d	60 d	180 d
Ref (RC)	15.6	15.4	10.8	8.8	6.8
8%MgO (MC)	15.4	15.2	10.7	8.4	6.7
0.5%Fiber (SFRC)	14.9	14.5	9.8	8.0	6.2
8%MgO + 0.5%Fiber (SFRMC-1)	14.8	14.0	8.6	7.7	5.1
8%MgO + 1%Fiber (SFRMC-2)	13.7	11.2	6.4	6.2	4.8
8%MgO + 1.5%Fiber (SFRMC-3)	11.1	10.3	7.2	6.2	4.9

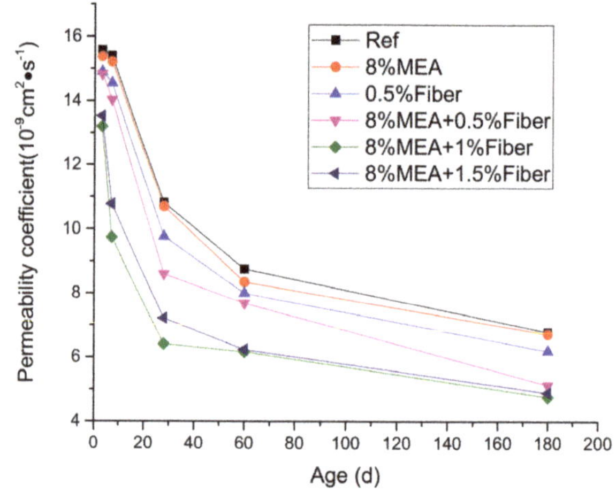

Figure 8. Changes of chloride diffusion coefficient.

Figure 8 shows that using MgO separately had little effect on CDC. At 180 days, CDC of MC was only 1.2% lower than that of RC. Different from MgO, steel fiber had both positive and negative effects on the durability of concrete. The first was the positive effect. Steel fiber could effectively inhibit the generation of cracks and reduce CDC. The other was the negative effect. The addition of steel fiber increased the number of the interface between fiber and matrix and provided a channel for the transfer of chloride ions, which possibly improved CDC. It can be found from Table 3, when the steel fiber was less than 1%, the positive factor played a major role. CDC decreased with the increase of steel fiber. Compared to RC, CDC of SFRC, SFRMC-1, SFRMC-2 decreased by 5.5%, 8.8%, and 36.7%, respectively at 7 days. However, when the steel fiber continued to increase, the negative effect gradually dominated, and CDC of SFRMC-3 was larger than that of SFRMC-2, which indicated that when steel fiber was too large, the negative effect caused by steel fiber was larger than the positive effect. Therefore, for SFRMC, the optimal content of steel fiber should not be more than 1%. In the later stage of curing (180 days), owing to the combined action of MgO and steel fiber, CDC of SFRMC-2 was 4.8×10^{-9} cm^2/s, which was only 70.8% of RC, and the durability of concrete was obviously enhanced.

3.4. Combined Action of Steel Fiber and MgO on Porosity of Concrete

As an important part of concrete, the pore structure directly affected the mechanical properties and durability of concrete [24–26]. Therefore, the study on the effect of MgO and steel fiber on concrete is inseparable from the research on the pore structure. The pores can be classified into four categories: cementitious pores (<10 nm), transitional pores (10–100 nm), capillary pores (100–1000 nm), and macro pores (>1000 nm) [27]. Figure 9 shows the pore diameter distribution curve of concrete. Figure 10 shows the statistics of pore size.

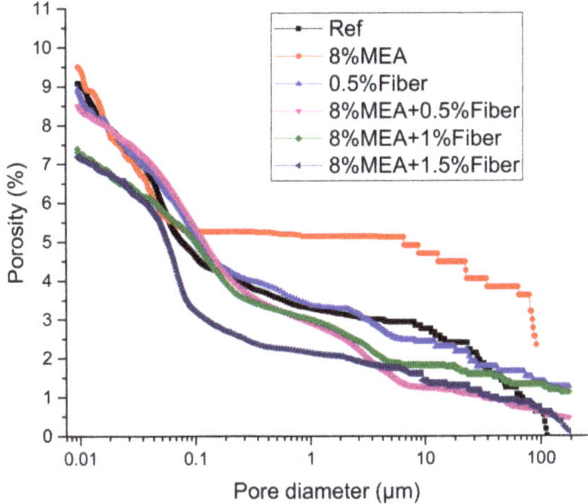

Figure 9. Porosity distribution curve of concrete.

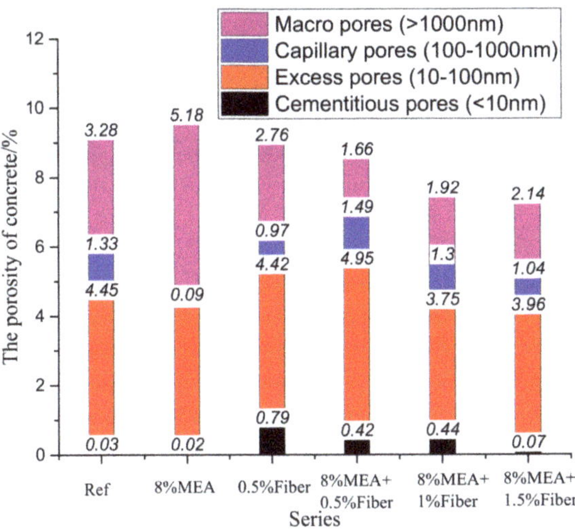

Figure 10. Statistics of concrete pore size.

Figures 9 and 10 show that compared to RC, the porosity of MC increased by 5.7%, indicating that the expansion of MgO is harmful to pore structure. The porosity of SFRC decreased by 1.5% compared to RC, and the number of large pores decreased by 15.9%, indicating that the steel fiber optimized the pore structure of concrete.

At the same time, when MgO and steel fiber were used together, the porosity continued to decrease. Compared to RC, the porosity of SFRMC-1, SFRMC-2, SFRMC-3 decreased by 6.2%, 18.4%, and 20.6%, respectively. In addition, the number of large pores was significantly reduced, cementitious and excessive pores were increased, and the connectivity of pores was reduced, which was also the main reason for the improvement of the chloride diffusion resistance of SFRMC.

3.5. Combined Action of Steel Fiber and MgO on Interfacial Transition Zone

To further analyze the principle of performance increase of SFRMC, the microstructure of concrete, especially the interfacial transition zone (ITZ) between fiber and matrix, was explored. The microstructure of ITZ with different mix ratios is shown in Figure 11. Figure 11 shows that in SFRC, there were obvious long and wide gaps in ITZ due to the shrinkage of matrix. Besides, tensile stress was produced because steel fiber restricted the shrinkage of matrix. In addition, when the tensile stress was larger than the tensile strength of concrete, the vertical crack was produced, leading to a reduction in durability.

Different from SFRC, when steel fiber and MgO were used at the same time, the wide gap at the interface disappeared. This change was mainly due to the extrusion and filling effect of MgO. Owing to the expansion under the constraint of steel fiber, self-stress was generated in ITZ (Figure 12). The self-stress made the biting force and friction force greatly increased, made ITZ become dense, and the boundary become fuzzy.

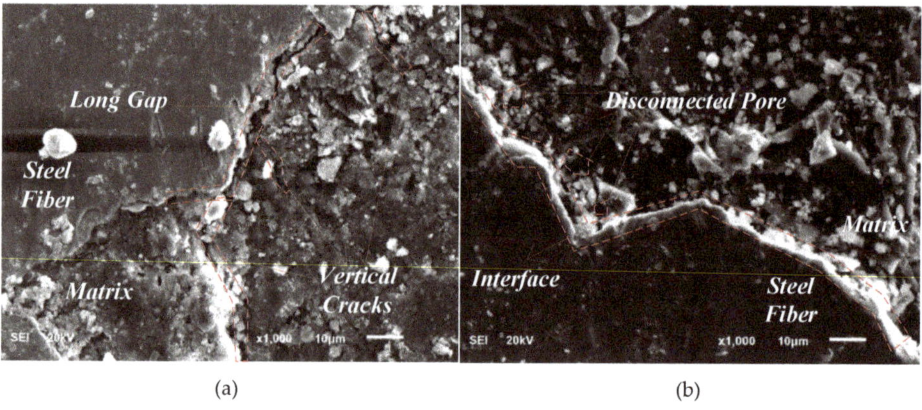

(a)   (b)

Figure 11. Interfacial transition zone (ITZ) between the steel fiber and matrix. (**a**) SFRC, (**b**) SFRMC-2.

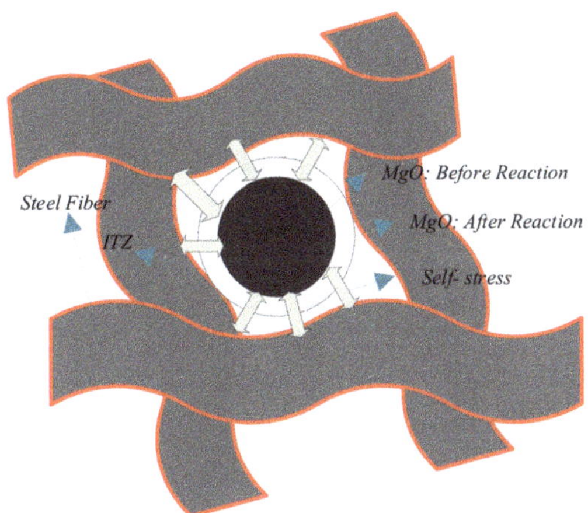

Figure 12. Self-stress caused by expansion extrusion at the interface.

4. Conclusions

Considering the defects of MgO and steel fiber when they were used separately, MgO and steel fiber were used at the same time in this paper. The influence of combined action on split tensile strength, chloride diffusion resistance, and pore structure of concrete was discussed in detail. Through detailed experimental research and theoretical analysis, the following conclusions could be drawn:

(1) For concrete, MgO and steel fiber could be used at the same time. In SFRMC, the steel fiber played the role of bearing and limiting expansion at the same time. The combination of MgO and steel fiber improved the split tensile strength and chloride diffusion resistance of concrete significantly. Compared to RC, the tensile strength of SFRMC-1, SFRMC-2, and SFRMC-3 increased by 3.1%, 61.3%, and 64.5%, and CDC decreased by 8.8%, 36.7%, and 33.1%, respectively.

(2) When steel fiber and MgO are used simultaneously, the two have a synergistic effect and the performance of the concrete is greatly improved. The combined action of steel fiber and MgO reduced the porosity of concrete, which could not be achieved by steel fiber and MgO separately. Compared to RC, the porosity of SFRMC-1, SFRMC-2, and SFRMC decreased by 6.2%, 18.4%, and 20.6%, respectively.

(3) Through the analysis of pore structure, the mechanism of performance enhancement of SFRMC was studied. Owing to the expansion of MgO was restrained by steel fiber, MgO changed from outward expansion to inward extrusion, resulting in filling and compaction effect. At the same time, self-stress was produced in ITZ, which improved the interfacial strength between steel fiber and matrix.

Author Contributions: Conceptualization, F.J. and M.D.; methodology, L.M.; software, F.J.; validation, F.J., M.D., L.M. and W.W.; formal analysis, M.D. and L.M.; investigation, F.J., M.D., L.M. and W.W.; resources, F.J. and M.D.; data curation, W.W.; writing—original draft preparation, F.J. and M.D.; writing—review and editing, F.J., M.D., L.M. and W.W.; visualization, F.J.; supervision, F.J., M.D., L.M. and W.W.; project administration, F.J.; funding acquisition, F.J. All authors have read and agreed to the published version of the manuscript.

Funding: This research was funded by Science and Technology Development Plan of Suzhou (SNG201904), National Key Research and Development Plan of China (2017YFB0309903-01) and Transportation Science and Technology Planning Project of Shandong Province (2018B37-02).

Acknowledgments: The authors would like to thank Zhongyang Mao from Nanjing Tech University for his precious contribution in the experiments.

Conflicts of Interest: The authors declare no conflict of interest.

References

1. Feng, J.; Sun, W.; Zhai, H.; Wang, L.; Dong, H.; Wu, Q. Experimental Study on Hybrid Effect Evaluation of Fiber Reinforced Concrete Subjected to Drop Weight Impacts. *Materials* **2018**, *11*, 2563. [CrossRef]
2. Amini, F.; Barkhordari Bafghi M, A.; Safayenikoo, H.; Sarkardeh, H. Strength of Different Fiber Reinforced Concrete in Marine Environment. *Mater. Sci.* **2018**, *24*, 204–211. [CrossRef]
3. Ma, K.; Qi, T.; Liu, H.; Wang, H. Shear Behavior of Hybrid Fiber Reinforced Concrete Deep Beams. *Materials (Basel, Switzerland)* **2018**, *11*, 2023. [CrossRef]
4. Yang, G.; Wei, J.; Yu, Q.; Huang, H.; Li, F. Investigation of the Match Relation between Steel Fiber and High-Strength Concrete Matrix in Reactive Powder Concrete. *Materials (Basel, Switzerland)* **2019**, *12*, 1751. [CrossRef]
5. Alsaif, A.; Bernal, S.A.; Guadagnini, M.; Pilakoutas, K. Durability of Steel Fibre Reinforced Rubberised Concrete Exposed to Chlorides. *Constr. Build. Mater.* **2018**, *188*, 130–142. [CrossRef]
6. Gao, D.; Zhang, L.; Zhao, J.; You, P. Durability of Steel Fibre-Reinforced Recycled Coarse Aggregate Concrete. *Constr. Build. Mater.* **2020**, *232*, 117119. [CrossRef]
7. Afroughsabet, V.; Biolzi, L.; Monteiro, P.J.M. The Effect of Steel and Polypropylene Fibers On the Chloride Diffusivity and Drying Shrinkage of High-Strength Concrete. *Composites Part B Engineering* **2018**, *139*, 84–96. [CrossRef]
8. Guo, Y.; Hu, X.; Lv, J. Experimental Study on the Resistance of Basalt Fibre-Reinforced Concrete to Chloride Penetration. *Constr. Build. Mater.* **2019**, *223*, 142–155. [CrossRef]
9. Wang, Y.; Zhang, S.; Niu, D.; Su, L.; Luo, D. Strength and Chloride Ion Distribution Brought by Aggregate of Basalt Fiber Reinforced Coral Aggregate Concrete. *Constr. Build. Mater.* **2020**, *234*, 117390. [CrossRef]
10. Ramli, M.; Kwan, W.H.; Abas, N.F. Strength and Durability of Coconut-Fiber-Reinforced Concrete in Aggressive Environments. *Constr. Build. Mater.* **2013**, *38*, 554–566. [CrossRef]
11. Ding, Y.; Bai, Y. Fracture Properties and Softening Curves of Steel Fiber-Reinforced Slag-Based Geopolymer Mortar and Concrete. *Materials (Basel, Switzerland)* **2018**, *11*, 1445. [CrossRef]
12. Li, F.; Cao, C.; Cui, Y.; Wu, P. Experimental Study of the Basic Mechanical Properties of Directionally Distributed Steel Fibre-Reinforced Concrete. *Adv. Mater. Sci. Eng.* **2018**, 1–11. [CrossRef]
13. Ige, O.; Barnett, S.; Chiverton, J.; Nassif, A.; Williams, J. Effects of Steel Fibre-Aggregate Interaction on Mechanical Behaviour of Steel Fibre Reinforced Concrete. *Adv. Appl. Ceram.* **2017**, *116*, 193–198. [CrossRef]
14. Xu, L.; Huang, C.; Liu, Y. Expansive Performance of Self-Stressing and Self-Compacting Concrete Confined with Steel Tube. *J. Wuhan Univ. Technology-Mater. Sci. Ed.* **2007**, *22*, 341–345. [CrossRef]
15. Hadi, M.N.S.; Al-Tikrite, A. Behaviour of Fibre-Reinforced RPC Columns Under Different Loading Conditions. *Constr. Build. Mater.* **2017**, *156*, 293–306. [CrossRef]
16. Choi, W.C.; Jung, K.Y.; Jang, S.J.; Yun, H.D. The Influence of Steel Fiber Tensile Strengths and Aspect Ratios on the Fracture Properties of High-Strength Concrete. *Materials (Basel)* **2019**, *12*, 2105. [CrossRef]
17. Mo, L.; Deng, M.; Tang, M.; Al-Tabbaa, A. MgO Expansive Cement and Concrete in China: Past, Present and Future. *Cement Concrete Res.* **2014**, *57*, 1–12. [CrossRef]
18. Polat, R.; Demirboğa, R.; Khushefati, W.H. Effects of Nano and Micro Size of CaO and MgO, Nano-Clay and Expanded Perlite Aggregate on the Autogenous Shrinkage of Mortar. *Constr. Build. Mater.* **2015**, *81*, 268–275. [CrossRef]
19. Mo, L.; Deng, M.; Tang, M. Effects of calcination condition on expansion property of MgO-type expansive agent used in cement-based materials. *Cem. Concr.Res.* **2010**, *40*, 437–446. [CrossRef]
20. Dung, N.T.; Unluer, C. Improving the Performance of Reactive MgO Cement-Based Concrete Mixes. *Constr. Build. Mater.* **2016**, *126*, 747–758. [CrossRef]
21. Jiang, F.; Mao, Z.; Deng, M.; Li, D. Deformation and Compressive Strength of Steel Fiber Reinforced MgO Concrete. *Materials* **2019**, *12*, 3617. [CrossRef] [PubMed]
22. MOHURD. *Standard for Test Method of Mechanical Properties of Ordinary Concrete, GB/T 50081-2016*; China Standards Press: Beijing, China, 2016.
23. ASTM. *Standard Test Method for Electrical Indication of Concrete's Ability to Resist Chloride Ion Penetration, C1202*; ASTM International: West Conshohocken, PA, USA, 2012.
24. Zhang, M.; Li, H. Pore Structure and Chloride Permeability of Concrete Containing Nano-Particles for Pavement. *Constr. Build. Mater.* **2011**, *25*, 608–616. [CrossRef]

25. Poon, C.S.; Kou, S.C.; Lam, L. Compressive Strength, Chloride Diffusivity and Pore Structure of High Performance Metakaolin and Silica Fume Concrete. *Constr. Build. Mater.* **2006**, *20*, 858–865. [CrossRef]
26. Güneyisi, E.; Gesoğlu, M.; Mermerdaş, K. Improving Strength, Drying Shrinkage, and Pore Structure of Concrete Using Metakaolin. *Mater. Struct.* **2008**, *41*, 937–949. [CrossRef]
27. Guo, J. The Theoretical Research of the Pore Structure and Strength of Concrete. Ph.D. Thesis, Zhejiang University, Hangzhou, China, 2004.

Article

Study of Bond Strength of Steel Bars in Basalt Fibre Reinforced High Performance Concrete

Piotr Smarzewski

Department of Structural Engineering, Faculty of Civil Engineering and Architecture,
Lublin University of Technology, 20-618 Lublin, Poland; p.smarzewski@pollub.pl; Tel.: +48-698-695-284

Received: 22 April 2020; Accepted: 28 May 2020; Published: 29 May 2020

Abstract: The paper presents the study on bond behaviour of steel bars. It reports the research conducted on local bond strength of short length specimens in high performance concrete (HPC) and basalt fibre reinforced high performance concrete (BFRHPC). In this study, the basalt fibre volume content, concrete cover, bar diameter and rib geometry are the main parameters. Further important factors are the directions of the casting and loading. Determining the effect of aforementioned main parameters on the bond strength in test series is required, in order to design reinforced HPC structures. The study of local bond strength in HPC and BFRHPC with five different basalt fibre fractions included tests of seventy-two short length specimens, using two concrete cover and two diameters of steel bars with different rib face angles. For different ranges of BFRHPC strength, relationships for bond strength with respect to the splitting tensile strength were obtained. The bond strength increased with the splitting tensile strength and compressive strength of BFRHPC specimens with the 12 mm and 16 mm bar respectively. The bond strength of BFRHPC was lower for the bar with the greater distances between the lugs on the bar.

Keywords: bond strength; high performance concrete; reinforcing steel bar; basalt fibre

---

1. Introduction

High performance concrete (HPC) has compressive strength above 80 MPa and low permeability. A serious disadvantage of this composite is brittleness, which increases with strength. Numerous researchers have revealed that steel and polypropylene fibres [1–6] or combinations of these fibres [7–10] can reduce the brittleness of HPC and significantly improve its tensile strength and fracture toughness, as well as its ductility.

Basalt fibres produced from molten basalt rock have very good strength properties, as well as high resistance to fire and alkaline environment, and at the same time are relatively cheap. These characteristics determine their use in concrete [11,12]. However, the cost in addition to the chemical and mechanical properties of basalt fibres vary, depending of the type and quality of the raw material and the production process of these fibres [13]. Nevertheless, the above-mentioned characteristics and an environmentally friendly manufacturing process [14] might determine their application in HPC structures, instead of the most commonly used steel and polypropylene fibres.

The research into basalt fibre reinforced concrete (BFRC) has largely been focused on fundamental mechanical properties, such as compressive, splitting tensile and flexural strength, as well as fracture toughness [11,12,14–18]. Fibre reinforced high performance concrete (FRHPC) is widely regarded as an excellent composite for use in sustainable construction [19–21]. However, optimum fibre dosages vary significantly in different types of concrete, such as geopolymeric concrete [5], normal strength concrete [14,17], high performance concrete [16,19–21] or ultra-high performance concrete [22–25]. Although the basalt fibre reinforced high performance concrete (BFRHPC) has good tensile strength, the rather poor ductility of this composite [19–21] means that concrete structures with basalt fibres should contain steel or other reinforcing bars to achieve the required performance and reliability. For this reason, a good interfacial bond of BFRHPC to steel bars is an important determinant of the interaction of these two materials in the structure.

Designing reinforced HPC structures with basalt fibres requires knowledge of the bond behaviour of steel reinforcement. Due to the high compressive strength, HPC elements have small cross-sections, and for this reason, the concrete cover and the failure mode are the most important parameters. When using fibres, knowledge about their impact on the bond behaviour and the possibility of replacing transverse reinforcement is very important [26]. In construction practice, the bar diameter, rib geometry, pouring direction of concrete and load direction are also relevant.

The bond strength between concrete and reinforcing bars in short lengths depends, among other factors, on the bar diameter, concrete strength and concrete cover. The theory of partly cracked thick cylinder proposed by Tepfers is the most comprehensive approach for determining the local bond, although it does not take into account the effect of deformation properties and the geometry of ribs in reinforcing bars [27]. This theory assumes that an uncracked concrete ring confines the cracked concrete and the reinforcing bar, and is resistant to bursting stresses radiating outwards from the bar at an angle of 45° to the bar axis. Soretz and Holzenbein [28] reported that the bonding action is different when the rib face angle is less than 30°, and they presented the dependence of the deformation pattern on the behaviour of local bond stress-slip as a function of the relative rib area. Darwin and Graham revealed that a bar with a smaller rib face angle results in lower bond strength [29]. Hwang et al. [30] determined the impact of silica fume on the splice strength of deformed bars embedded in high strength concrete (HSC) and noted that the bond strength of a beam at 10% cement replacement by silica fume was 15% lower than the beam bond strength with no silica fume addition. Tests carried out on beams with a long embedded length showed that the bond stresses differ significantly along the length [31]. Esfahani and Rangan [32] studied the bond strength in HPC concrete and estimated the maximum bond stress based on the results of short length specimens in which the distribution of the bond stresses at failure was almost uniform. Local bond equations were proposed in this investigation that can be used to determine the maximum bond stress at the ends of the splices. Holschemacher et al. [33] showed that the high brittleness of the composite does not adversely affect the bond behaviour of steel bars anchored in ultra-high strength concrete. The binding stiffness increased, due to the high elastic modulus and compressive strength of concrete. Eligehausen et al. [34] examined different failure modes of anchoring bars (pull-out, pry-out and splitting) in normal strength concrete. It was found that the various failure modes were influenced by different parameters, such as confinement, relative rib area, fibre addition, concrete cover and casting direction. Alkaysi and El-Tawil [35] performed pull-out tests to characterize the bond strength of a non-proprietary ultra-high performance concrete mixture, and revealed that the bond strength decreases with increased embedment length and increases with higher volume content of steel fibres.

2. Research Significance

Designing reinforced concrete structures is relied on the bond relationship between the concrete and the steel bar. However, investigations of the interfacial bond behaviour between BFRHPC and steel bars have not yet been published, and remain a pressing need. The experimental program described in this article fills this gap and aimed to quantify the bond strength between HPC and steel reinforcement for several important design parameters. To assess the effect of different parameters on the bond, the maximum bond stress was required, which was estimated from the test results of short length specimens and an almost uniform distribution of the bond stresses at failure. A total of 72 bar pull-out tests were performed with parameters including average bar coatings (43 mm, 93 mm), nominal bar diameter (12 mm, 16 mm) and fibre volume content (0%, 1%, 1.25%, 1.5%, 1.75% and 2%). The present study is part of an investigation on the bond strength in high and ultra-high performance concrete with the application of hybrid fibres. Several relationships for the bond strength with respect to the splitting tensile strength and fibre volume fractions of BFRHPC were suggested. In addition, the equations obtained in this study for determining the local bond strength can be used as the maximum bond stress at the ends of the splices.

3. Materials and Methods

HPC was made with general use Portland cement CEM I 52.5R (C) (CEMEX, Chełm, Poland), non-densified silica fume (SF) (Ironworks Łaziska, Łaziska Górne, Poland), tap water (W) and well-graded coarse and fine aggregates (CA, FA). The physical properties and chemical compositions of cementitious materials are summarized in Table 1.

Table 1. Chemical compositions and physical properties of used cement and silica fume.

Composition (%)	Cement	Silica Fume
SiO_2	19.99	85.0
Al_2O_3	4.19	-
Fe_2O_3	3.76	-
CaO	64.82	1.0
MgO	1.14	-
SO_3	3.25	2.0
K_2O	0.46	-
Na_2O	0.24	3.0
Cl	0.07	0.3
Si	-	0.4
Loss on ignition	3.01	4.0
Insoluble matter	0.18	-
Specific surface area (cm^2/g)	4839	150,000
Water demand (%)	30	-
Start of setting (min)	160	-
End of setting (min)	210	-
Compressive strength at 2 days (MPa)	40.3	-
Tensile strength at 2 days (MPa)	6.5	-

Superplasticizer CX ISOFLEX 793 (CEMEX Admixtures GmbH, Salzkotten, Germany) (Sp) based on polycarboxylate ethers was used in high dosage. The 12 mm lengths of chopped basalt fibers (Holtex, Rzgów, Poland) (BF) were used. The basalt fibre bundles are flat, approximately 1 mm wide and made of 13 μm diameter filaments. Other characteristics of the fibre were as follows: density 2.7 g/cm^3, modulus of elasticity 70 GPa, tensile strength 1700 MPa and elongation at break 2.5%. The fibres used in this study are shown in Figure 1a. The ribbed bars, on which the bond strength was determined, were B500SP with diameters of 12 mm and 16 mm. Axial tensile tests were carried out on three specimens for each diameter of reinforcing bars (Figure 1b).

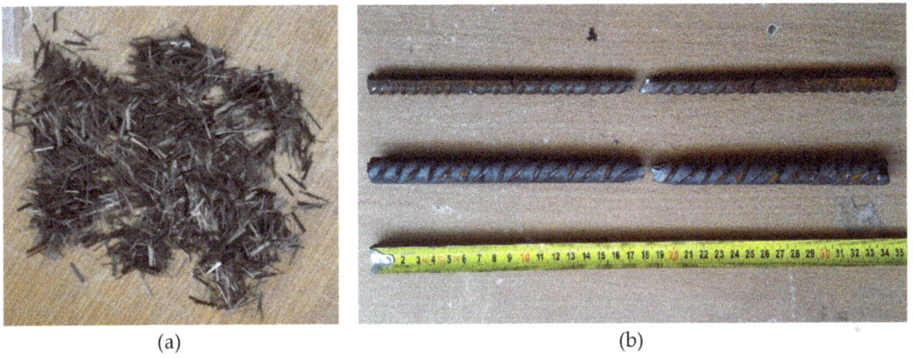

| (a) | (b) |

Figure 1. (a) Fibres used in experimental work; (b) reinforcing bars after tensile test.

The measured yield strength, tensile strength, elastic modulus, ultimate tensile strain and minimum relative rib area of 12 mm and 16 mm were 605 MPa, 691 MPa, 203 GPa, 33%, 0.040 mm^2 and 622 MPa, 716 MPa, 211 GPa, 27%, 0.056 mm^2 respectively. In order to measure the rib geometries, the reinforcing bars were longitudinally sliced. The rib face angle in the 12 mm and 16 mm bars was between 37° and 73°, as well as 44° and 73°, correspondingly. The measured distances between the ribs for the above-mentioned bars were from 3 to 6 mm and from 6 to 12 mm, respectively.

HPC specimens used in this investigation were cast with a 0.28 water-binder ratio. The silica fume had a specific surface of 15 m^2/g. The basalt coarse aggregate (CA) had a maximum size of 5 mm, fineness modulus of 5.92 and compressive strength of 196 MPa. The fine quartz sand aggregate (FA) had a maximum particle size of 2 mm and fineness modulus of 1.84. The quantities used in the reference mixture were as follows: cement—670.5 kg/m^3, silica fume—74.5 kg/m^3, coarse aggregate—990 kg/m^3, fine aggregate—500 kg/m^3, water—210 L/m^3, and superplasticizer—20 L/m^3. The reference mixture HPC-B0 did not contain any basalt fibres. The following five contained fibres with an aspect ratio of $l/d = 923$ and percentage ranging from 1% to 2% were made with a reduced quartz sand amount equal to the weight of the added fibres. A summary of the all mixture types in this work is shown in Table 2.

Table 2. Test mixture proportions.

Designation	C (kg/m^3)	SF (kg/m^3)	CA (kg/m^3)	FA (kg/m^3)	W (L/m^3)	Sp (L/m^3)	BF (kg/m^3)	BF (%)
HPC–B0	670.5	74.5	990	500	210	20	0	0
HPC–B1	670.5	74.5	990	473	210	20	27	1
HPC–B1.25	670.5	74.5	990	466.25	210	20	33.75	1.25
HPC–B1.5	670.5	74.5	990	459.5	210	20	40.5	1.5
HPC–B1.75	670.5	74.5	990	452.75	210	20	47.25	1.75
HPC–B2	670.5	74.5	990	446	210	20	54	2

Two different types of deformed bars with nominal diameters of 12 mm and 16 mm were used in pull-out tested specimens. Seventy-two specimens in six series were produced and examined. All the specimens were cast parallel to the steel bar fixed vertically in the mould according to the pre-adopted position of the bar. Thirty-six specimens were made for each bar using six different basalt fibre contents of 0%, 1%, 1.25%, 1.5%, 1.75% and 2%. There were three specimens for each combination of c_y/ϕ_b and fibre content. The ratios of the side cover to the bottom cover of the concrete (c_x/c_y) were 1 and 2.1. The details of the pull-out test specimens and test set-up are shown in Table 3 and Figure 2.

Table 3. Details of test specimens.

Bar Diameter ϕ_b (mm)	Bottom Concrete Cover, c_y (mm)	Side Concrete Cover, c_x (mm)	Embedded Length, l_e (mm)	c_y/ϕ_b	Number of Specimens
12	94	94	60	7.83	18
12	44	94	60	3.67	18
16	92	92	80	5.75	18
16	42	92	80	2.63	18

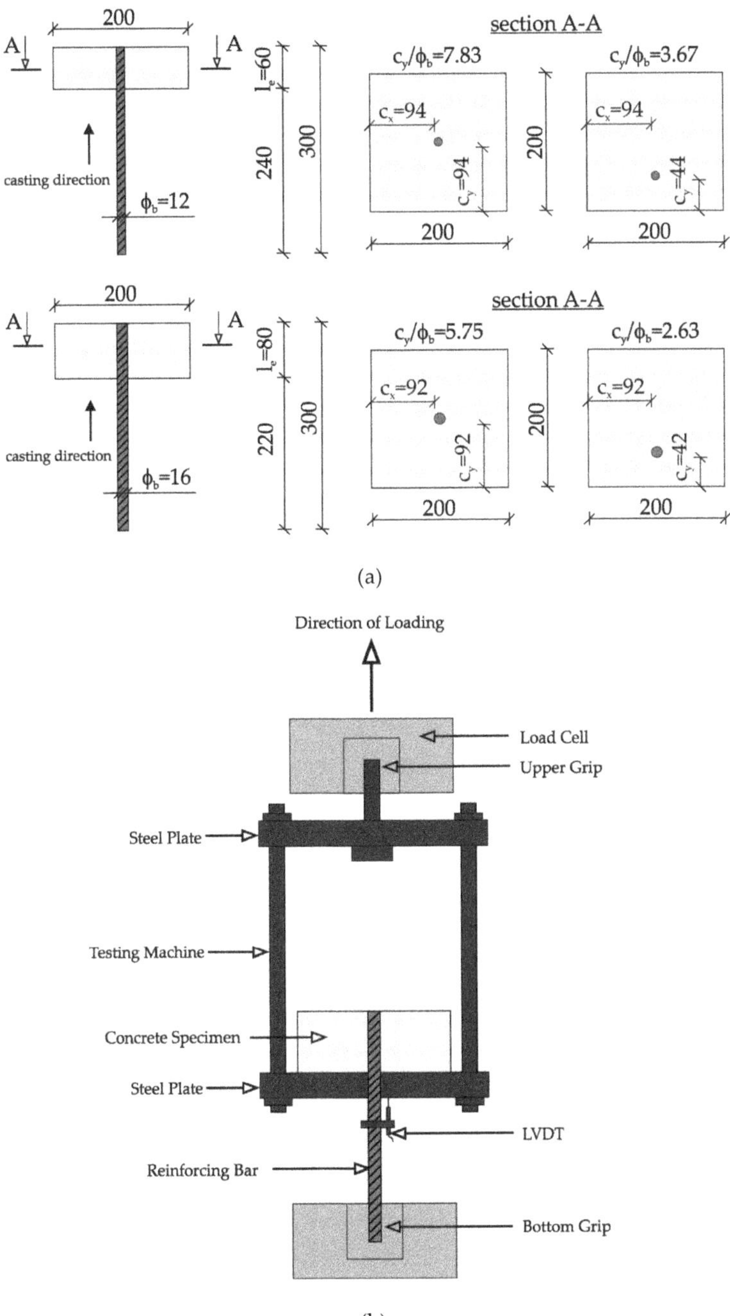

Figure 2. (**a**) Test specimen details (dimensions in mm), (**b**) test set-up.

To investigate the bond behaviour of steel bars embedded in HPC and BFRHPC, a pull-out test was carried out. A pull-out load was applied using the MTS 319.25 servo-hydraulic testing machine

(MTS, Eden Prairie, MN, USA) with a maximum flexural load capacity of 250 kN under displacement control, with a rate of 1 mm/min during testing. The LVDT was used to measure bond slip between the reinforcing bar and the HPC/BFRHPC at the loaded end. In all specimens, longitudinal cracking of the concrete cover occurred over the anchorage length of the reinforcing bars before failure. All the specimens failed due to splitting of concrete.

For each casting, seventy-two 100 mm × 100 mm × 100 mm HPC cubes were made to determine the compressive and splitting tensile strength. These strength tests were carried out using an Advantest 9 load-controlled universal press (CONTROLS, Milan, Italy) of 3 MN capacity. The average of six measurements was recorded as the compressive and splitting tensile strength of each HPC. Eighteen 100 mm × 100 mm × 500 mm beam specimens were made to set the flexural strength. The bending tests were subjected to three-point loading using the MTS 319.25 press (MTS, Eden Prairie, MN, USA). The beam specimens were supported on two rolls spaced at a distance of 300 mm and then were loaded at the midspan. The flexural tests used deflection as the control signal at a rate of 0.05 mm/min. The average of the flexural strength of three beam specimens was reported for each HPC.

4. Results and Discussion

4.1. Hardened State Properties

Table 4 gives the results of the compressive strength, splitting tensile strength and flexural strength for all HPC at 28 days, with standard deviations (SD) and coefficients of variation. The ratio of plain HPC strength to basalt fibre reinforced HPC strength are also presented.

Table 4. Mechanical properties of high performance concretes (HPCs).

Property	HPC–B0	HPC–B1	HPC–B1.25	HPC–B1.5	HPC–B1.75	HPC–2
Compressive strength (MPa)	135.5	112.1	112.3	116.4	111.1	105.3
SD (MPa)	1.85	1.31	1.31	1.17	1.38	1.21
CV (%)	1.36	1.17	1.17	1.00	1.24	1.15
Ratio	1.00	0.83	0.83	0.85	0.82	0.78
Splitting tensile strength (MPa)	6.4	8.0	9.0	9.6	7.9	7.7
SD (MPa)	0.15	0.21	0.29	0.19	0.11	0.23
CV (%)	2.34	2.62	3.22	1.98	1.39	2.99
Ratio	1.00	1.25	1.41	1.50	1.23	1.20
Flexural strength (MPa)	6.0	10.1	10.8	12.4	11.5	12.2
SD (MPa)	0.15	0.23	0.39	0.29	0.21	0.33
CV (%)	2.50	2.28	3.61	2.34	1.83	2.70
Ratio	1.00	1.68	1.80	2.07	1.92	2.03

The results indicate that the basalt fibre volume content influenced the compressive strength in fibre content range considered from 0% to 2%. The compressive strength of HPC was reduced by 15–22%, with an increase in the fibre volume fraction with comparison with that of HPC without fibres. This can be clarified by higher porosity in the fibre reinforced concrete due to the air infiltration during the mixing procedure of the fibres and concrete [36]. Additionally, it was observed that basalt fibres can absorb water from the cement paste, and reduce the HPC fluidity and compressive strength [19]. Although the addition of basalt fibres resulted in a decrease in compressive strength, they noticeably enhanced the toughness of the HPC under compression. The literature also suggests that the primary benefit of basalt fibres in concrete under compression is the change from a brittle failure mode to a more ductile mode [11,14,16,17].

It can be noticed that the tensile splitting and flexural strength of BFRHPC is always greater than these of HPC for the reason that basalt fibres act as crack-arrestors. Such an action may be assigned to the basalt fibres role in gradually filling micro-cracks, which leads to increased bonding in high performance concrete microstructure, and, consequently, to higher splitting tensile strength and flexural strength.

The addition of basalt fibre to normal strength concrete can significantly increase its splitting tensile and flexural strength. Jiang et al. [11] revealed that basalt fibres with 12 mm length and between 0.30% and 0.5% by volume content increased the flexural strength and splitting tensile strength of normal strength concrete by 9.58% and 24.34%, respectively. Kabay [37] obtained 10.3% increase in flexural strength when 2 kg/m^3 and 4 kg/m^3 basalt fibres were added to normal strength concrete. Çelik and Bingöl [38] stated that the addition of basalt fibre at 0.20% volume content increase flexural strength of self-compacting concrete 11.58%. On the other hand, the highest splitting tensile strength was observed at 0.30% basalt fibre content, which increased by 12.78% according to the plain specimen.

The maximum values of split tensile and flexural strength were reached for HPC with 1.5% basalt fibre fraction. It was found that its tensile strengths compared to reference mixture increased by 50% and 106.67%, respectively. The decreases in average split tensile and flexural strength of HPC with 1.75% and 2% basalt fibre contents may be related to the variable distribution of fibres and the tendency to group them in these mixtures. Çelik and Bingöl [38] also observed a reduction in the flexural strength of self-compacting concrete with an increase in the addition of basalt fibres from 0.2% to 0.25%. They found that may be due to the fact that the low flexibility of basalt fibres caused the formation of larger voids in the concrete and deteriorated the distribution of fibres.

Concerning the tensile strength results, the flexural strength exhibited more variability and much higher results than splitting strength. Considering the CV values given in Table 4, it can be observed that the addition of basalt fibres did not increase the error indicators.

Çelik and Bingöl [38] studied normal strength concrete with basalt fibres with a volume fraction of 0.30% at 28 days. Based on micrographs analyses, it was noticed that there are voids between basalt fibres and cement paste, which reduce fibre-matrix bond and, as a consequence, rather brittle post-cracking behaviour occurs in basalt fibre reinforced composites. Branston et al. [14] also found that the density of cement on the basalt fibre surface decreases considerably after nine months. However, high performance concrete is characterised by a dense interfacial transition zone (ITZ), and behaves differently to normal strength concrete. Wu et al. [39] performed microstructure analyses of the steel fibre in ultra-high performance concrete matrix, calculated changes in porosity with the distance from the edge of the fibre and found that the porosity decreased as the distance from the edge of the fibre increased. The porosity around the fibre at 50 μm was 47% lower than the porosity at 10 μm from the fibre edge. On this basis, it can be assumed that the porosity of high performance concrete with basalt fibres will also grow with increasing fibre content, and smaller distances between the fibres, which consequently results in a decrease in HPC strength.

4.2. Bond Strength Versus Fibre Content Relationship

The average values of bond properties are summarized in Table 5. The bond strength was calculated as the achieved pull-out force divided by the initial surface area of the embedded bar by means of the relation $T_b = F_{max}/\pi\phi_b l_e$ where F_{max} is the maximum force in the bar at bond failure in the specimen. It is important to note that splitting failure mode did represent a bond failure in all the HPC specimens.

Table 5 shows the increase in nominal and normalized bond strength with increasing basalt fibre volume content up to 1.5%. This is caused by that since the deformed steel bar provides bonding through interlocking between the rebar lugs and the surrounding concrete, the bond strength increases with increasing strength of the surrounding high performance concrete [40]. In addition, the average displacement at peak load increased when the basalt fibre content increased, similar to the bond strength. The average normalized bond strength from all tested series was obtained as 1.2. The fibre content had a significant impact on this value. It can be seen that the normalized bond strength was twice as high in concrete with 2% fibre content, as compared to plain HPC.

Figure 3 compares the bond strength versus the fibre content relationships using two different c_y/ϕ_b relations for two bar diameters ϕ_b, namely 12 mm and 16 mm. The error indicator denotes the standard deviation.

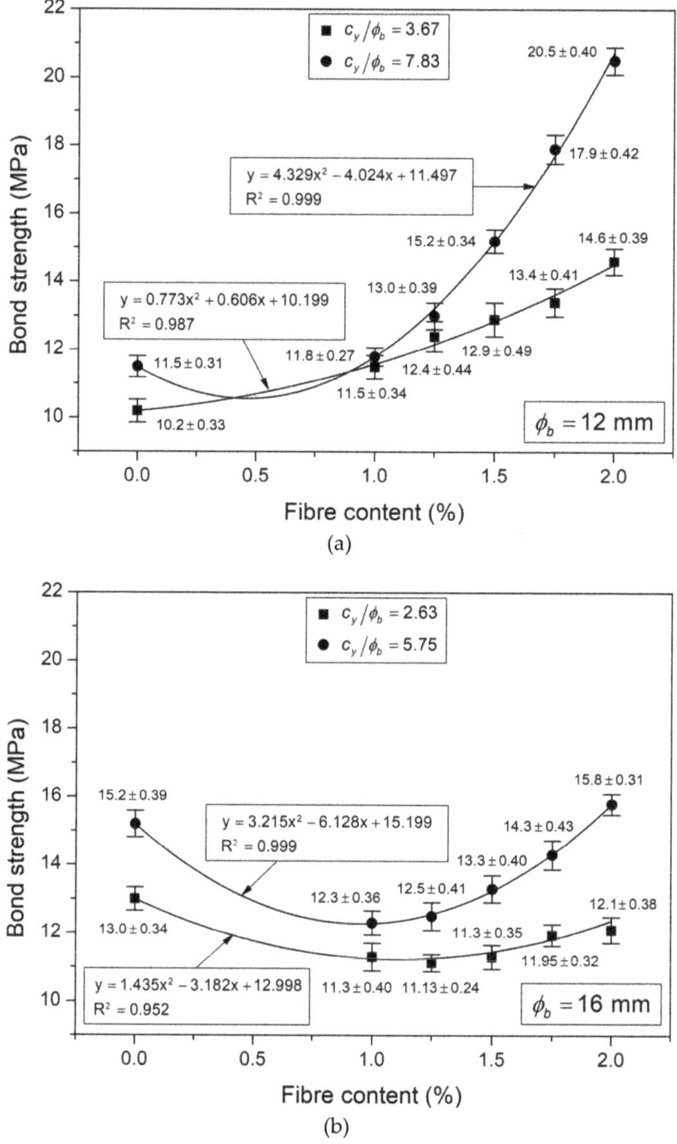

Figure 3. Variation in bond strength with fibre content for different bar diameters ϕ_b and relations c_y/ϕ_b (a) 12 mm, (b) 16 mm.

The comparison of the results in Figure 3a shows that, for different values of c_y/ϕ_b, the bond strengths of the 12 mm bar increase with a growing basalt fibre content, and with an increasing concrete cover. The bond strengths of HPC-B0 were 11.5 MPa with the central position of the bar in the specimen and 10.5 MPa with the bar positioned halfway between the centre and the edge of the specimen and of BFRHPC provided an improvement at each volume fraction. The bond strength improvement of the 12 mm bar located axially in the specimens and pulled-out of BFRHPC ranged from 2.6% to 78.3% at the volume fractions from 1.0% to 2.0%, as well as form 12.7% to 43.1% for the eccentric pulled-out bar from BFRHPC at the same fractions. In contrast, the bond strength of the 16 mm bar falls when the fibre content is 1–1.25% (Figure 3b). The decreases ranged between 6.9% and 13.1% for an eccentric bar and reduced with higher fibre content. A similar downward trend in the range of 5.9–19.1% was maintained for the centrally located bar. Only for the 2% fibre volume fraction a slight 3.9% increase in the bond strength was noted. In the case of concrete without fibres, the bond strength was greater for the 16 mm bar in the range of 24.3–27.4%. This can be explained by the larger rib face angle between 44° to 73° of the 16 mm bar than in the 12 mm bar with a rib face angle from 37° to 73°. The lower values of bond strengths of BFRHPC for the 16 mm bar can be clarified by the larger distances between the ribs from 6 to 12 mm, compared to the 12 mm bar, in which these distances are between 3 to 6 mm. At larger distances between the lugs on the bar, the short basalt fibres could be oriented parallel to the bar between the ribs causing faster bond failure occurring, as a result of shearing off and crushing the concrete. This effect was observed for the smallest concrete cover. This indicates that the bond strength of the steel bar depends on the HPC strength. The BFRHPC compressive strength used in this study was 15–22% lower, compared to the concrete without fibres, and the splitting tensile strength was 20–50% higher.

Table 5. Summary of pull-out test results.

Designation	V_f (%)	f_c (MPa)	ϕ_b (mm)	l_e (mm)	c_y (mm)	c_x (mm)	F_{max} (kN)	τ_b (MPa)	δ_m (mm)	τ_b^* (MPa)
HPC–B0	0	135.5	12	60	94	94	26.01	11.5	1.66	0.99
					44	94	23.07	10.2	2.16	0.88
			16	80	92	92	61.12	15.2	3.77	1.31
					42	92	52.28	13.0	3.58	1.12
HPC–B1	1.0	112.1	12	60	94	94	26.69	11.8	3.52	1.11
					44	94	26.01	11.5	4.47	1.09
			16	80	92	92	49.46	12.3	4.53	1.16
					42	92	45.44	11.3	5.67	1.07
HPC–B1.25	1.25	112.3	12	60	94	94	29.41	13.0	5.90	1.23
					44	94	28.05	12.4	6.56	1.17
			16	80	92	92	50.27	12.5	5.62	1.18
					42	92	44.64	11.1	5.25	1.05
HPC–B1.5	1.5	116.4	12	60	94	94	34.38	15.2	5.95	1.41
					44	94	29.18	12.9	8.33	1.20
			16	80	92	92	53.48	13.3	6.77	1.23
					42	92	45.44	11.3	6.31	1.05
HPC–B1.75	1.75	111.1	12	60	94	94	40.49	17.9	5.99	1.70
					44	94	30.31	13.4	7.32	1.27
			16	80	92	92	57.50	14.3	5.69	1.36
					42	92	48.25	12.0	6.72	1.14
HPC–B2	2.0	105.3	12	60	94	94	46.37	20.5	6.23	2.00
					44	94	33.02	14.6	4.24	1.42
			16	80	92	92	63.54	15.8	7.89	1.54
					42	92	48.66	12.1	6.73	1.18

V_f = volume content of fibre, f_c = mean value of compressive strength, ϕ_b = bar diameter, l_e = embedded length, c_y, c_x = bottom and side concrete cover, F_{max} = maximum pull-out load, τ_b = bond strength, δ_m = mean value of displacement at peak load, τ_b^* = normalized bond strength, $\tau_b^* = \tau_b / \sqrt{f_c}$.

4.3. Bond Strength Versus Splitting Tensile Strength Relationship

Figure 4 presents the bond strength versus splitting tensile strength relations for test series, using two different types of bars within a limited range of basalt fibre percentages.

The error indicators denote the standard deviations both the bond strength (0.24–0.49 MPa) and splitting tensile strength (0.15–0.29 MPa). The given relationships were proposed for basalt fibre content between 0% and 1.5%, due to problems with proper fibre distribution at 1.75% and 2% fibre volume contents, which resulted in a significant reduction of 18.8% in splitting tensile strength of BFRHPC. As can be seen in Figure 4, the values of BFRHPC bond strength for test with the 16 mm bar test are higher in the range of 4.2–32.2% than the bond strength in the 12 mm bar for the 0–1% fibre volume fraction. On the other hand, with fibre contents of 1.25–1.5%, it can be seen that the 12 mm bars were characterized by higher bond strengths in the range of 4–14.3%. It can be reported that the bond strength for the 12 mm bar increases when the BFRHPC splitting tensile strength growths. In contrast, the bond strength for the specimens with the 16 mm bar is higher when the concrete compressive strength growing (see Table 4). This is probably due to different geometry of the ribs in the 16 mm bar, as well as a different orientation of the fibres around the larger diameter bar, which may cause local sliding surfaces resulting in less values of the bond strength. Therefore, the extent of the concrete crushing in front of the ribs decreased as the compressive concrete strength increased.

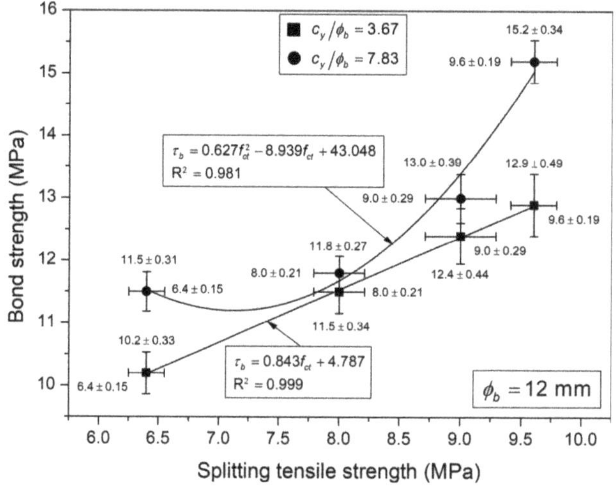

(a)

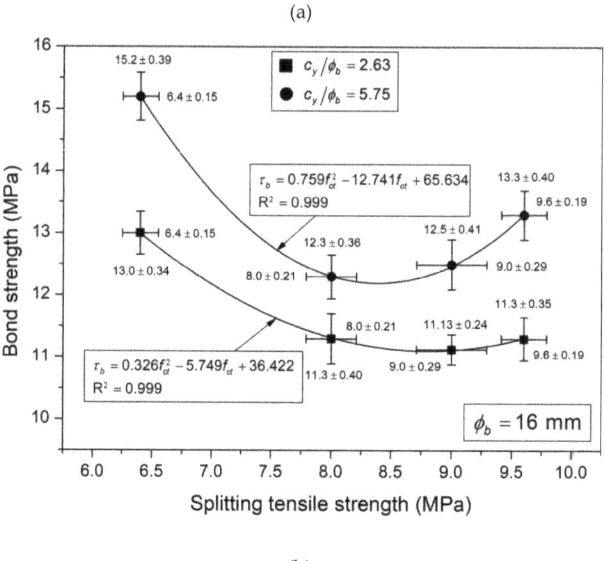

(b)

Figure 4. Variation in bond strength with splitting tensile strength for different bar diameters ϕ_b and relations c_y/ϕ_b (**a**) 12 mm, (**b**) 16 mm.

4.4. Bond Strength Versus Ratio of Concrete Cover to Bar Diameter Relationship

The values for the bond strength of BFRHPC are shown in Figure 5 as a function of c_y/ϕ_b ratio. Linear regression leads to the relationships of the concrete cover to bar diameter ratio versus bond strength given in Table 6. The test data for two different bars are grouped together in the analysis.

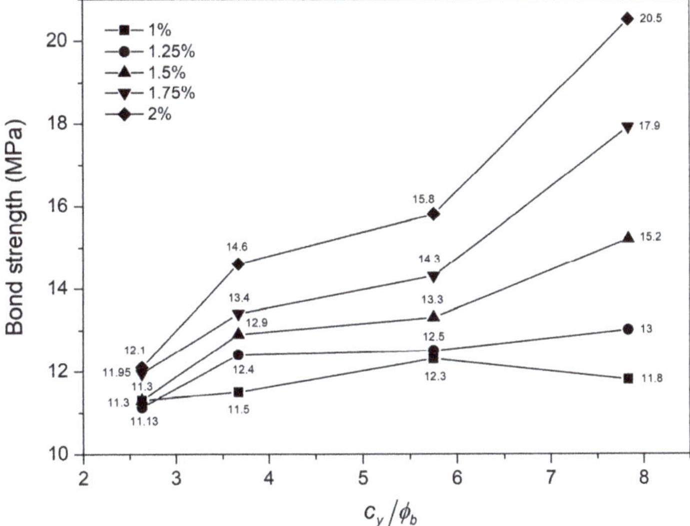

Figure 5. Variation in bond strength with c_y/ϕ_b ratio for different basalt fibre volume content.

Table 6. Ratio of concrete cover to bar diameter—bond strength relationship of basalt fibre reinforced high performance concrete (BFRHPC).

Designation	Equation	Reduced Chi-Square	Adj. R-Square
HPC–B1	$\tau_b = 0.122\frac{c_y}{\phi_b} + 11.117$	0.165	0.420
HPC–B1.25	$\tau_b = 0.299\frac{c_y}{\phi_b} + 10.773$	0.239	0.749
HPC–B1.5	$\tau_b = 0.667\frac{c_y}{\phi_b} + 9.862$	0.310	0.920
HPC–B1.75	$\tau_b = 1.063\frac{c_y}{\phi_b} + 9.102$	0.611	0.937
HPC–B2	$\tau_b = 1.483\frac{c_y}{\phi_b} + 8.379$	1.060	0.971

In general, the bond strength values for the BFRHPC specimens improved with increasing the basalt fibre content and c_y/ϕ_b ratio. For the 12 mm bars and c_y/ϕ_b ratios of 3.67 and 7.83, the bond strength increased by 27% and 73.7%, respectively, when the fibre volume content increased from 1% to 2%. For the 16 mm bars and c_y/ϕ_b ratios of 2.63 and 5.75, specimens containing 1% basalt fibres showed, in sequence, 6.6% and 22.1% lower bond strengths than specimens with 2% fibres. This seems to confirm the regularity that the bond strength depends on the quantity of fibres available to bridge any cracks formed under loading. Comparing the average bond strengths obtained for the 12 mm and 16 mm embedded bars, while maintaining a similar concrete cover and 2% fibre content, it was observed that the bond strength is up to 29.7% higher for a bar with a smaller diameter. Similar trends were noted for the remaining fibre volume fractions 1–1.75%. It can be also concluded that suggested linear relations agree very well with the test results. The recorded bond strength values differed most at the lowest fibre content of 1% and 1.25%, especially at greater concrete cover, which justifies further research.

5. Conclusions

The objective of this study was to investigate the bond strength between basalt fibre reinforced high performance concrete and deformed steel bar. Simple pull-out tests were carried out at two different bar diameters, two concrete covers, two embedment lengths and six fibre volume contents. Seventy-two short length pull-out specimens were tested at 28 days. Based on this investigation, the following main conclusions were drawn:

- The compressive strength of HPC worsened with increasing basalt fibres volume fraction. The compressive strength ranged from 15% to 22% lower for the fractions from 1% to 2%.
- The splitting tensile strength and flexural strength of BFRHPC significantly improved with the addition of basalt fibres at various volume contents. The splitting tensile and flexural strength showed a maximum at 1.5% fibre content but slight decreases at 1.75% and 2% contents, compared to 1.5% still remaining 23% and 20% higher, as well as 92% and 103% higher than before the fibre's addition.
- Changes in basalt fibre content between 1% and 2% resulted in differences between 7% and 74% in the bond strength achieved. Similar differences in bond strength were observed after normalization, suggesting that the bond strength depends more on the quantity of fibres available to bridge any cracks forming under load, rather than the differences in compressive strength of BFRHPC.
- The bond strength for the same bar diameter and fibre volume content improved from 3% to 40% with a two-fold concrete cover, suggesting that the bond strength is dependent on the fibre orientation in the concrete cover area.
- The relationships obtained for BFRHPC predict the bond strengths accurately.

Funding: This research was supported financially by the Polish Ministry of Science and Higher Education within statutory research project no. FN15/ILT/2020.

Acknowledgments: The financial support from the Polish Ministry of Science and Higher Education is greatly appreciated. The author would also like to thank the CEMEX Company for donating the cement and superplasticizer for this research.

Conflicts of Interest: The author declares no conflict of interest.

References

1. Sivakumar, A.; Santhanam, M. Mechanical properties of high strength concrete reinforced with metallic and non-metallic fibres. *Cem. Concr. Compos.* **2007**, *29*, 603–608. [CrossRef]
2. Smarzewski, P. Hybrid Fibres as Shear Reinforcement in High-Performance Concrete Beams with and without Openings. *Appl. Sci.* **2018**, *8*, 2070. [CrossRef]
3. Smarzewski, P. Analysis of Failure Mechanics in Hybrid Fibre-Reinforced High-Performance Concrete Deep Beams with and without Openings. *Materials* **2019**, *12*, 101. [CrossRef] [PubMed]
4. Smarzewski, P. Processes of Cracking and Crushing in Hybrid Fibre Reinforced High-Performance Concrete Slabs. *Processes* **2019**, *7*, 49. [CrossRef]
5. Afroughsabet, V.; Ozbakkaloglu, T. Mechanical and durability properties of high-strength concrete containing steel and polypropylene fibers. *Constr. Build. Mater.* **2015**, *94*, 73–82. [CrossRef]
6. Song, P.S.; Hwang, S. Mechanical properties of high-strength steel fiber-reinforced concrete. *Constr. Build. Mater.* **2004**, *18*, 669–673. [CrossRef]
7. Job, T.; Ramaswamy, A. Mechanical properties of steel fiber-reinforced concrete. *J. Mater. Civ. Eng.* **2007**, *19*, 385–392.
8. Ramezanianpour, A.A.; Ghahari, S.A.; Khazaei, A. Feasibility Study on Production and Sustainability of Poly Propylene Fiber Reinforced Concrete Ties Based on a Value Engineering Survey. In Proceedings of the Third International Conference on Sustainable Construction Materials and Technologies (SCMT3'13), Kyoto, Japan, 18–21 August 2013; pp. 1–8.

9. Shah, A.A.; Ribakov, Y. Recent trends in steel fibered high-strength concrete. *Mater. Des.* **2011**, *32*, 4122–4151. [CrossRef]

10. Feng, J.; Sun, W.; Zhai, H.; Wang, L.; Dong, H.; Wu, Q. Experimental study on hybrid effect evaluation of fiber reinforced concrete subjected to drop weight impacts. *Materials* **2018**, *11*, 2563. [CrossRef]

11. Jiang, C.; Fan, K.; Wu, F.; Chen, D. Experimental study on the mechanical properties and microstructure of chopped basalt fibre reinforced concrete. *Mater. Des.* **2014**, *58*, 187–193. [CrossRef]

12. High, C.; Seliem, H.M.; Adel El-Safty, A.; Rizkalla, S.H. Use of basalt fibers for concrete structures. *Constr. Build. Mater.* **2015**, *96*, 37–46. [CrossRef]

13. Fiore, V.; Scalici, T.; Di Bella, G.; Valenza, A. A review on basalt fibre and its composites. *Compos. Part B Eng.* **2015**, *74*, 74–94. [CrossRef]

14. Branston, J.; Das, S.; Kenno, S.Y.; Taylor, C. Mechanical behaviour of basalt fibre reinforced concrete. *Constr. Build. Mater.* **2016**, *124*, 878–886. [CrossRef]

15. Dias, D.P.; Thaumaturgo, C. Fracture toughness of geopolymeric concretes reinforced with basalt fibers. *Cem. Concr. Compos.* **2005**, *27*, 49–54. [CrossRef]

16. Ayub, T.; Shafiq, N.; Nuruddin, M.F. Mechanical properties of high-performance concrete reinforced with basalt fibers. *Procedia Eng.* **2014**, *77*, 131–139. [CrossRef]

17. Iyer, P.; Kenno, S.Y.; Das, S. Mechanical properties of fiber-reinforced concrete made with basalt filament fibers. *J. Mater. Civ. Eng.* **2015**, *11*, 04015015. [CrossRef]

18. Lipatov, Y.V.; Gutnikov, S.; Manylov, M.; Zhukovskaya, E.; Lazoryak, B. High alkali-resistant basalt fiber for reinforcing concrete. *Mater. Des.* **2015**, *73*, 60–66. [CrossRef]

19. Smarzewski, P. Flexural Toughness of High-Performance Concrete with Basalt and Polypropylene Short Fibres. *Adv. Civ. Eng.* **2018**, *2018*, 1–8. [CrossRef]

20. Smarzewski, P. Influence of basalt-polypropylene fibres on fracture properties of high performance concrete. *Compos. Struct.* **2019**, *209*, 23–33. [CrossRef]

21. Smarzewski, P. Flexural toughness evaluation of basalt fibre reinforced HPC beams with and without initial notch. *Compos. Struct.* **2020**, *235*, 111769. [CrossRef]

22. Yoo, D.-Y.; Shin, H.; Yang, J.-M.; Yoon, Y.-S. Material and bond properties of ultra high performance fiber reinforced concrete with micro steel fibers. *Compos. Part B Eng.* **2014**, *58*, 122–133. [CrossRef]

23. Smarzewski, P.; Barnat-Hunek, D. Fracture properties of plain and steel-polypropylene-fiber-reinforced high-performance concrete. *Mater. Tehnol.* **2015**, *49*, 563–571. [CrossRef]

24. Smarzewski, P. Effect of Curing Period on Properties of Steel and Polypropylene Fibre Reinforced Ultra-High Performance Concrete. *IOP Conf. Ser. Mater. Sci. Eng.* **2017**, *245*, 32059. [CrossRef]

25. Smarzewski, P. Study of Toughness and Macro/Micro-Crack Development of Fibre-Reinforced Ultra-High Performance Concrete After Exposure to Elevated Temperature. *Materials* **2019**, *12*, 1210. [CrossRef]

26. Fehling, E.; Lorenz, P.; Leutbeche, T. Experimental Investigations on Anchorage of Rebars in UHPC. In Proceedings of the International Symposium on Ultra High Performance Concrete, Kassel, Germany, 7–9 March 2012.

27. Tepfers, R. *A Theory of Bond Applied to Overlapped Tensile Reinforcement Splices of Deformed Bars*; Report 73-2; Chalmers University of Technology: Göteborg, Sweden, 1973.

28. Soretz, S.; Holzenbein, H. Influence of Rib Dimensions of Reinforcing Bars on Bond and Bendability. *ACI J.* **1979**, *76*, 111–126.

29. Darwin, D.; Graham, E.K. Effect of Deformation Height Spacing on Bond Strength of Reinforcing Bars. *ACI Struct. J.* **1993**, *90*, 646–657.

30. Hwang, S.J.; Lee, Y.Y.; Lee, C.S. Effect of Silica Fume Splice Strength of Deformed Bars of High-Performance Concrete. *ACI Struct. J.* **1994**, *91*, 294–302.

31. Tepfers, R. Bond stress along lapped reinforcing bars. *Mag. Concr. Res.* **1980**, *32*, 135–142. [CrossRef]

32. Local Bond Strength of Reinforcing Bars in Normal Strength and High-Strength Concrete (HSC). *ACI Struct. J.* **1998**, *95*, 96–106. [CrossRef]

33. Holschemacher, K.; Weiße, D.; Klotz, S. Bond of Reinforcement in Ultra High Strength Concrete. In Proceedings of the International Symposium on Ultra High Performance Concrete, Kassel, Germany, 13–15 September 2004.

34. Eligehausen, R.; Mallée, R.; Silva, J.F. *Anchorage in Concrete Construction*; Ernst&Sohn: Berlin, Germany, 2006.

35. Alkaysi, M.; El-Tawil, S. Factors affecting bond development between Ultra High Performance Concrete (UHPC) and steel bar reinforcement. *Constr. Build. Mater.* **2017**, *144*, 412–422. [CrossRef]

36. Wang, D.; Ju, Y.; Shen, H.; Xu, L. Mechanical properties of high performance concrete reinforced with basalt fiber and polypropylene fiber. *Constr. Build. Mater.* **2019**, *197*, 464–473. [CrossRef]

37. Kabay, N. Abrasion resistance and fracture energy of concretes with basalt fiber. *Constr. Build. Mater.* **2014**, *50*, 95–101. [CrossRef]

38. Çelik, Z.; Bingöl, A.F. Mechanical properties and postcracking behavior of self-compacting fiber reinforced concrete. *Struct. Concr.* **2019**, 1–10. [CrossRef]

39. Wu, Z.; Khayat, K.H.; Shi, C. How do fiber shape and matrix composition affect fiber pullout behavior and flexural properties of UHPC? *Cem. Concr. Compos.* **2018**, *90*. [CrossRef]

40. Harajli, M.; Hamad, B.; Karam, K. Bond-slip response of reinforcing bars embedded in plain and fiber concrete. *J. Mater. Civ. Eng.* **2002**, *14*, 503–511. [CrossRef]

Article

Inhibition of the Alkali-Carbonate Reaction Using Fly Ash and the Underlying Mechanism

Xin Ren, Wei Li, Zhongyang Mao and Min Deng *

College of Materials Science and Engineering, Nanjing Tech University. Nanjing 210009, China;
201761100171@njtech.edu.cn (X.R.); 201762100015@njtech.edu.cn (W.L.); mzy@njtech.edu.cn (Z.M.)
* Correspondence: dengmin@njtech.edu.cn; Tel.: +86-159-5059-1582

Received: 10 May 2020; Accepted: 3 June 2020; Published: 5 June 2020

Abstract: In this paper, fly ash is used to inhibit the alkali-carbonate reaction (ACR). The experimental results suggest that when the alkali equivalent (equivalent Na_2Oeq) of the cement is 1.0%, the adding of 30% fly ash can significantly inhibit the expansion in low-reactivity aggregates. For moderately reactive aggregates, the expansion rate can also be reduced by adding 30% of fly ash. According to a polarizing microscope analysis, the cracks are expansion cracks mainly due to the ACR. The main mechanisms of fly ash inhibiting the ACR are that it refines the pore structure of the cement paste, and that the alkali migration rate in the curing solution to the interior of the concrete microbars is reduced. As the content of fly ash increases, the concentrations of K^+ and Na^+ and the pH value in the pore solution gradually decrease. This makes the ACR in the rocks slower, such that the cracks are reduced, and the expansion due to the ACR is inhibited.

Keywords: inhibit; alkali-carbonate reaction; fly ash; expansion; mechanism

1. Introduction

In 1940, Stanton [1] first discovered the alkali-aggregate reaction (AAR), which has attracted the attention of many researchers. As it is extremely destructive and difficult to repair, it has been called the "cancer" of concrete. The AAR is divided into alkali-silicate reactions (ASRs) and alkali-carbonate reactions (ACRs). The ACR is one of the main problems in the long-term durability of concrete.

According to the time sequence, the expansion mechanism of the ACR can be divided into three types. Firstly, Gillott [2] believed that the expansion is the result of an increase in the solid volume due to water absorption by the clay, where dolomitization only provides a way for clay to absorb water. Secondly, Tang and Tong [3–5] believed that, although the absolute volume of the solid phase of the alkali-dolomite reaction is reduced in theory, the rearrangement and crystallization of the reaction products in a restricted space causes the expansion and cracking of the aggregate, respectively, leading to concrete cracking. Thirdly, Katayama [6–8] believed that the ACR is the combination of harmful expansion caused by the ASR of microcrystalline quartz and harmless dolomitization. Dolomitization produces the brucite and carbonate reaction ring, but the expansion is caused by the ASR due to microcrystalline quartz. However, Chen, X [9] and Chen, B [10] used tetramethylammonium hydroxide (TMAH) as the curing solution to investigate the expansion characteristics caused only by the ACR, as TMAH does not react with SiO_2. Their results showed that the ACR exists separately and contributes to the expansion. Huan Yuan [11], who also used TMAH as the curing solution, proved (by an expansion stress test and concrete microbar expansion test) that the alkali-carbonate reaction causes expansion.

The concrete durability problem caused by the AAR can be controlled by many methods, such as reducing the available alkali and adding a suitable amount of ash or chemical additives [12]. Supplementary cementitious materials (SCMS) have different effects in reducing the expansion due to the ACR and ASR [13]. It is well-known that supplementary cementitious materials have a significant effect in inhibiting the ASR [14]. Based on the success of these materials, they have been used to prevent the deterioration caused by the ACR. According to the research of Alireza Joshaghani [15], fly ash and trass can reduce the expansion rate of the ACR. At 56 d, using 10%, 20%, and 30% fly ash can reduce the expansion rate of mortar bars by 47%, 95%, and 73%, respectively, according to the ASTM C1260 standard. According to the experimental results of concrete prisms, the inhibitory effect of trass on the ACR is slightly better than that of fly ash. In a long-term test, the optimal content of fly ash was 20% and the optimal content of trass was 30%. Shehata's [16] experiment demonstrated that the expansion of concrete prisms mixed with SCMs exceeded the threshold of 0.040% in two years and no SCM had a complete effect on the ACR in the long-term, although some types were more effective than others at reducing expansion.

Shehata [16] and Min Deng [17] have shown that using fly ash to effectively inhibit the ACR expansion of the highly reactive dolomite limestones from Kingston, Canada is difficult. However, whether the ACR of low-reactive dolomite rocks and moderately reactive dolomite rocks can be effectively inhibited has not yet been studied. At present, the related specifications stipulate that concrete works cannot use the ACR reactive rocks as aggregates, limiting the application of dolomite rocks. Research on fly ash on the ACR of low-reactive dolomite rocks and moderately reactive dolomite rocks is therefore necessary. This serves to play a guiding role in the engineering application of dolomitic aggregates.

The main purpose of this study is to examine the inhibition and mechanism of fly ash on the ACR. Although the impact of fly ash on the ASR has been well-documented, there has been little research on the impact of fly ash on the ACR. In addition, most of the studies have focused on highly reactive aggregates, such as the Kingston dolomite limestone of Canada, with no clear research on low-reactive and moderately reactive rocks. In this study, therefore, we focus on low- and moderately reactive dolomite rocks.

2. Materials and Methods

2.1. Materials

The used materials were (1) low-alkali Portland cement (Type II) obtained from the Jiangnan Cement Plant, Nanjing, China, with 0.54% equivalent Na_2Oeq; (2) class F fly ash (FA) obtained from Henan; and (3) dolomitic limestones obtained from Baofuling Mountain, Shandong, China (BFL8) and Shuijingwan, Guizhou, China (SJW). The chemical compositions of the cement and fly ash are presented in Table 1. The chemical compositions of the dolomitic limestones are shown in Table 2.

Table 1. Chemical analysis of cement and fly ash (wt. %).

Samples	Chemical Compositions (wt. %)								
	SiO_2	CaO	MgO	Al_2O_3	Fe_2O_3	SO_3	K_2O	Na_2O	LOI
Cement (Type II)	22.02	60.51	2.18	6.34	3.05	1.86	0.47	0.23	1.96
Fly Ash	48.91	5.01	1.03	34.18	5.22	1.20	0.89	0.39	1.05

Table 2. Chemical analysis of aggregates (wt. %).

Samples	Chemical Compositions (wt. %)					
	SiO_2	CaO	MgO	Al_2O_3	Fe_2O_3	LOI
BFL8	2.68	48.65	4.39	0.93	0.26	42.06
SJW	7.68	42.68	5.21	1.26	0.88	38.64

The microscopic appearances of the fly ash and cement observed with a scanning electron microscope (SEM; JSM-6510, JEOL Co, Tokyo, Japan) are shown in Figure 1. The fly ash is shown in (A) and the cement is shown in (B). We can see, from Figure 1, that the fly ash particles were rounded and most particles in the cement were irregularly shaped. The X-ray diffraction (XRD; SmartLab X-ray diffractometer, Rigaku Co., Tokyo, Japan) analysis results are shown in Figure 2. The XRD analysis of SJW is shown in (A) and BFL8 is shown in (B). SJW rocks mainly include calcite, quartz, dolomite and muscovite. However, BFL8 rocks only have calcite, quartz and dolomite.

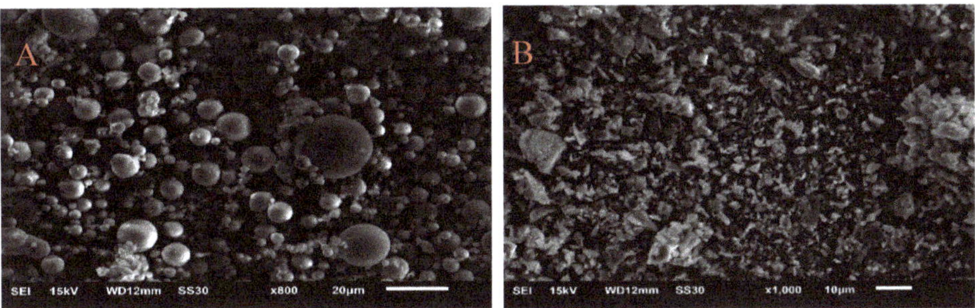

Figure 1. Scanning electron microscope images of (**A**) fly ash and (**B**) cement.

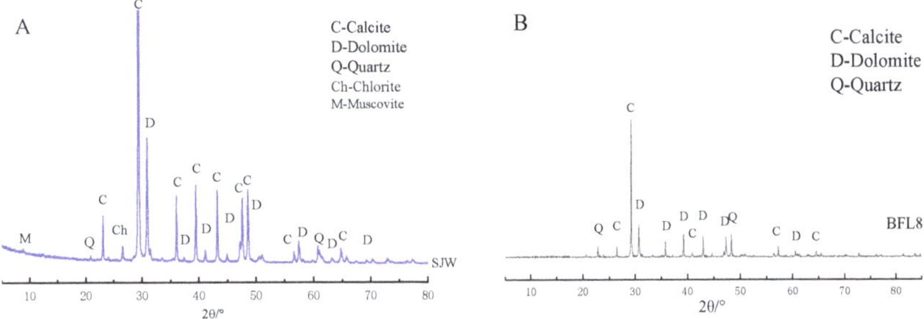

Figure 2. XRD analysis of (**A**) SJW and (**B**) BFL8.

2.2. Methods

2.2.1. Mortar Bars Test

A set of mortar bars ($25 \times 25 \times 285$ mm) were prepared with cement as well as dolomitic rocks with grain sizes of 0.125–4 mm, according to RILEM AAR-2 [18]. The alkali equivalent (equivalent Na_2Oeq) of the cement was adjusted to 0.9%, the ratio of aggregate to cement was 1:1, and the water–cement ratio was adjusted to 0.32. The designs of the mortar bars, according to RILEM AAR-2, are shown in Table 3. All specimens were kept for 24 h in a moisturized condition for curing before demolding. Then, samples were transferred to 1 mol/L NaOH solution immediately and stored in sealed containers at 80 °C. When curing to a set age, remove the sample and measure the length, then calculate the expansion rate according to Equation (1):

$$P_t = (L_t - L_0)/(L_0 - 2b) \times 100\% \tag{1}$$

where P_t is the expansion rate after t days of curing, in %; L_t is the test piece length after t days of curing, in mm; L_0 is the initial length of the test piece, in mm; and b is the length of the nail embedded in the concrete, in mm.

Table 3. Mortar bars mix designs according to RILEM AAR-2.

Sample Name	w/c	Cement/g	Aggregates/g
BFL8-1	0.47	400	900
SJW-1	0.47	400	900

2.2.2. Concrete Microbars Test

A set of concrete microbars ($40 \times 40 \times 160$ mm) were prepared with cement, as well as dolomitic rocks with grain sizes of 5–10 mm, according to RILEM AAR-5 [19]. The alkali equivalent (equivalent Na_2Oeq) of the cement was adjusted to 1.5%. The ratio of aggregate to cement was 1:1, and the water–cement ratio was adjusted to 0.32. The designs of the microbars, according to RILEM AAR-5, are shown in Table 4. All specimens were kept for 24 h in a moisturized condition for curing before demolding. Then, samples were transferred to 1 mol/L NaOH solution immediately and stored in sealed containers at 80 °C. The expansion rate is calculated according to Equation (1).

Table 4. Concrete microbars mix designs according to RILEM AAR-5.

Sample Name	w/c	Cement/g	Fly Ash/g	Aggregates/g
BFL8-2-FA0%	0.32	900	0	900
BFL8-2-FA10%	0.32	810	90	900
BFL8-2-FA20%	0.32	720	180	900
BFL8-2-FA30%	0.32	630	270	900
SJW-2-FA0%	0.32	900	0	900
SJW-2-FA10%	0.32	810	90	900
SJW-2-FA20%	0.32	720	180	900
SJW-2-FA30%	0.32	630	270	900

2.2.3. Concrete Prisms Test

A set of concrete prisms were prepared with cement, as well as dolomitic rocks with grain sizes of 4.75–19 mm, according to ASTM C1293 [20]. The alkali equivalent (equivalent Na_2Oeq) of the cement was adjusted to 1.25%. The water–cement ratio was adjusted to 0.45. The designs of the concrete prisms, according to ASTM C1293, are shown in Table 5. All specimens were kept for 24 h in a moisturized condition for curing before demolding. Then, samples were transferred to 100% humidity at 38 °C. The expansion rate is calculated according to Equation (1).

Table 5. Concrete prisms mix designs according to ASTM C1293.

Sample Name	w/c	Cement/g	Fly Ash/g	Aggregates/g	Sand/g
BFL8-3-FA0%	0.45	3360	0	6000	3980
BFL8-3-FA30%	0.45	2352	1008	6000	3980
SJW-3-FA0%	0.45	3360	0	6000	3980
SJW-3-FA30%	0.45	2352	1008	6000	3980

2.2.4. Measurement of the PH, Ionic Concentration and Pore Structure

Three sets of samples ($20 \times 20 \times 20$ mm) were prepared with cement and fly ash, where the cement was replaced with varying amounts of fly ash. The samples were separately cured in a 1 mol/L NaOH solution and saturated $Ca(OH)_2$ solution at 80 °C. First, to determine the curing age, 5.0 g of cement paste was weighed. Second, the contents were transferred to a covered dish and broken up. Third, enough water was added to make up a volume of 200 ml. Fourth, the mixture was incubated at room temperature for 1 h, and stirred frequently. Then, the mixture was filtered with medium-quality filter paper into a 500 ml volumetric flask and washed with hot water. Finally, the K^+ and Na^+ concentrations of the pore solution were measured with a flame photometer (FP650, Aopu Co.

Shanghai, China) and the pH value of the pore solution was measured with a pH meter (PHS-25, INESA Co. Shanghai, China).

Three sets of samples ($20 \times 20 \times 20$ mm) were prepared with cement and fly ash. The water–cement ratio was adjusted to 0.32. They were separately cured in 1 mol/L NaOH solution at 80 °C. To determine the curing age, the cement paste was taken and the porosity and pore size distribution were measured using mercury intrusion porosimetry (MIP, AutoPore W 9500, Micromeritics Co. New York, NY, USA). As shown in Table 6, four mix formulations of cement pastes were designed to comparatively investigate the influences of fly ash on the pH, ionic concentration, and pore structure of the cement pastes.

Table 6. Mix proportions of the cement pastes.

Sample Name	w/c	Cement/g	Fly Ash/g
FA-0%	0.32	450	0
FA-10%	0.32	405	45
FA-20%	0.32	360	90
FA-30%	0.32	315	135

2.2.5. Thin Section Petrography

Samples taken from the concrete microrods were cut into thin slices. Then, the slices were examined with an optical microscope to check for the presence of reaction products and expansion cracks derived from the aggregate. A polarized optical microscope with transmitted light (Optiphot-II Pol reflectometer, × 25–400, Nikon, Tokyo, Japan) and a stereomicroscope (BH-2, Nikon, Tokyo, Japan) were used for this purpose. The preparation of thin sections was carried out according to the "Thin section specimen preparation" section in [21].

3. Results and Discussion

3.1. Discrimination of the Aggregate Alkali Reactive

The results of the mortar bar test (according to RILEM AAR-2) and concrete microbar test (according to RILEM AAR–5) are shown in Figures 3 and 4, respectively.

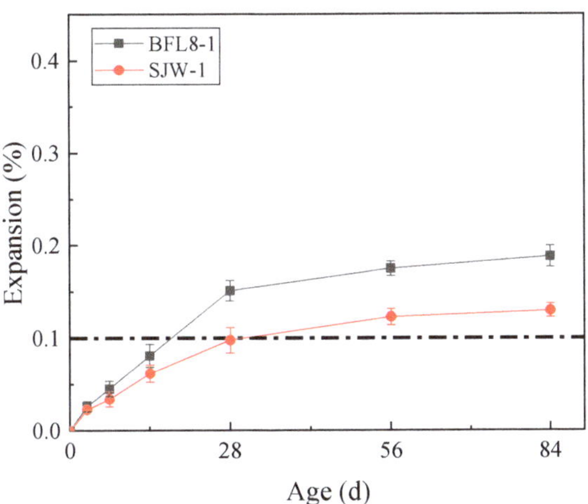

Figure 3. The expansion of the mortar bars prepared according to RILEM AAR-2.

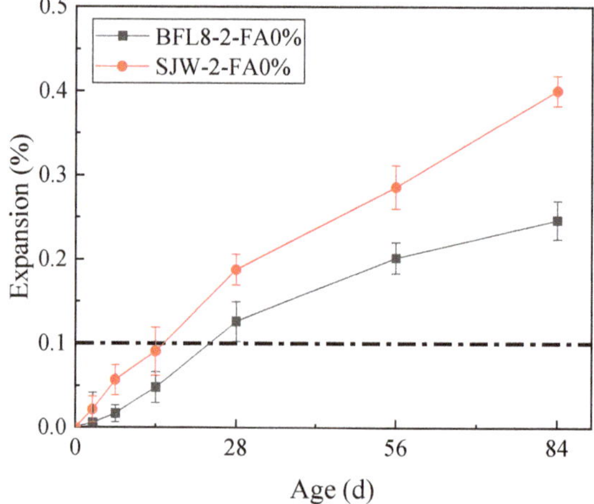

Figure 4. The expansion of the concrete microbars prepared according to RILEM AAR-5.

Figure 3 shows that the expansion of BFL8 and SJW at 28 d was 0.126% and 0.188%, respectively. At 28 d, their expansion was greater than 0.1%, which is indicated by the position of the dotted line in the figure. According to RILEM AAR-5, the aggregates of BFL8 and SJW both demonstrated alkali-carbonate reactions.

Figure 4 shows that the expansion of BFL8 and SJW at 14 days was 0.081%, 0.062%, respectively. The dotted line in Figure 4 is the threshold value for judging whether the aggregate demonstrates an alkali–silicate reaction. At 14 d, the expansions of BFL8 and SJW were less than 0.1%; therefore, according to RILEM AAR-2, the BFL8 and SJW aggregates had no alkali-silicate reactions

Based on Figures 3 and 4, we can see from the expansion rate that BFL8 has low reactivity and SJW has moderate reactivity. Furthermore, BFL8 and SJW only have alkali-carbonate reactions.

3.2. The Effect of Fly Ash on Concrete Microbars in the Short Term

3.2.1. Alkali Equivalent (Equivalent Na_2Oeq) of the Cement Adjusted to 1.5%

Figure 5 shows the expansion results of the 180-day concrete microbar test conducted on samples using different concentrations of fly ash, according to RILEM AAR-5, where the alkali equivalent (equivalent Na_2Oeq) of the cement was adjusted to 1.5%. The expansion of concrete microbars prepared by BFL8 is shown in (A) and the expansion of concrete microbars prepared by SJW is shown in (B).

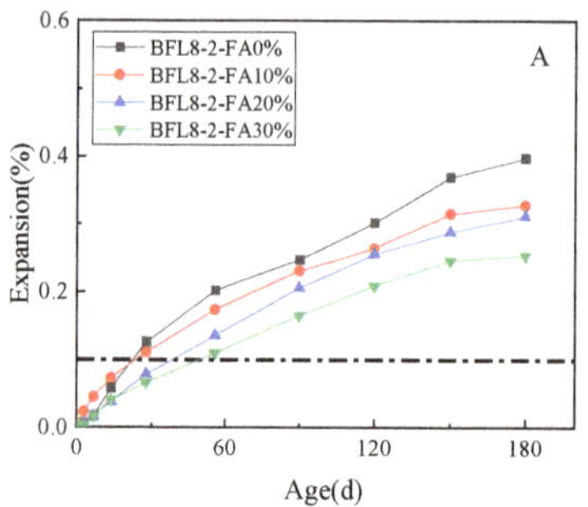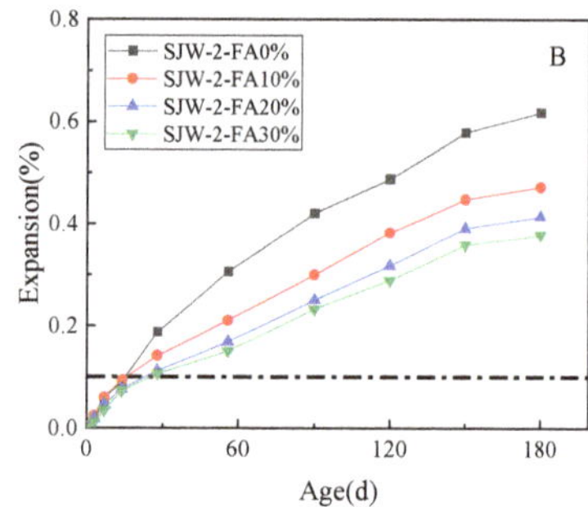

Figure 5. The expansion of the concrete microbars with fly ash: (**A**) concrete microbars prepared by BFL8 and (**B**) concrete microbars prepared by SJW.

Figure 5 shows that the expansion rates of BFL8 and SJW without fly ash were 0.126% and 0.188%, respectively, at 28 days. The BFL and SJW samples expanded about 0.398% and 0.618%, respectively, at 180 days. Compared with the concrete microbars mixed with 0% fly ash, the expansion of the concrete microbars mixed with 10%, 20%, and 30% fly ash and prepared with BFL8 decreased by 18%, 42%, and 51%, respectively, at 28 d, and by 17 %, 21%, and 36%, respectively, at 180 d. The expansion of the concrete microbars mixed with 10%, 20%, and 30% fly ash and made of SJW decreased by 29%, 43%, and 46%, respectively, at 28 d, and by 23%, 33%, and 39%, respectively, at 180 d. After curing in 80 °C 1 mol/L NaOH solution for 56 days, the expansion of the concrete microbars prepared with BFL8 mixed with 10%–30% fly ash was greater than 0.10%. After curing for 28 days in 1mol/L NaOH solution, the expansion of the concrete microbars prepared with SJW mixed with 10%–30% fly ash was greater than 0.10%. From this point of view, when the alkali equivalent (equivalent Na_2Oeq) of the cement is 1.5%, 10%–30% fly ash cannot effectively inhibit the expansion of the low-reactivity BFL8 and the moderately reactive SJW.

3.2.2. Alkali Equivalent (Equivalent Na_2Oeq) of the Cement Adjusted to 1.0%

When the alkali equivalent (equivalent Na_2Oeq) of the cement was 1.5%, fly ash could not effectively inhibit the ACR. Therefore, we reduced the alkali equivalent (equivalent Na_2Oeq) of the cement to study the effect when the alkali equivalent (equivalent Na_2Oeq) of the cement was 1.0%, in order to determine whether fly ash can effectively inhibit the expansion caused by the ACR under such conditions.

Figure 6 shows the expansion results of a 240-day concrete microbar test conducted on samples using different concentrations of fly ash according to RILEM AAR-5, where the alkali equivalent (equivalent Na_2Oeq) of the cement was 1.0%. The expansion of concrete microbars prepared by BFL8 is shown in (A) and the expansion of concrete microbars prepared by SJW is shown in (B).

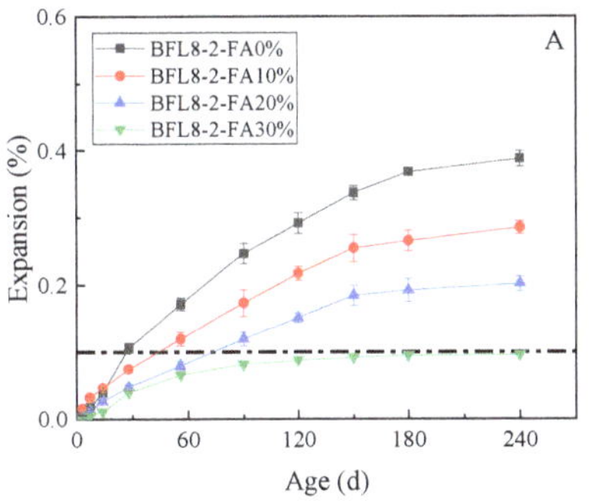

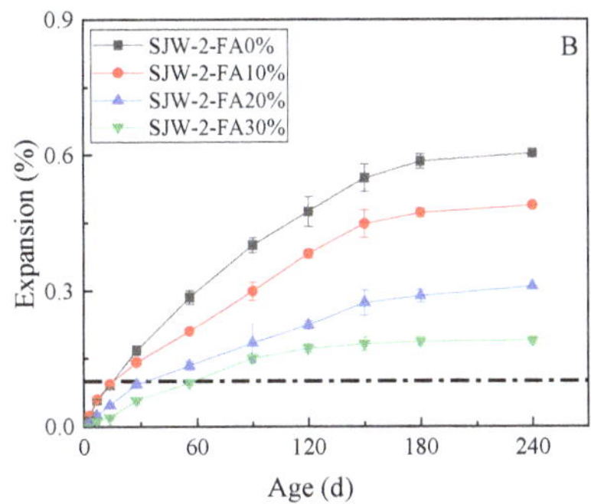

Figure 6. The expansion of the concrete microbars with fly ash: (**A**) concrete microbars prepared by BFL8 and (**B**) concrete microbars prepared by SJW.

Figure 6 shows that the expansion rates of BFL8 and SJW without fly ash were 0.106% and 0.168%, respectively, at 28 days. These samples of BFL8 and SJW expanded by approximately 0.388% and 0.603%, respectively, at 240 days, and their expansion basically stabilized. As the content of fly ash increased, the expansion of the concrete microbars made of BFL8 and SJW gradually decreased. At 28 d and 240 d, the addition of 10%, 20%, and 30% fly ash reduced the expansion of the concrete microbars prepared with BFL8 by 35%–66% and 26%–75%, respectively. Furthermore, adding 10%, 20%, and 30% fly ash reduced the expansion of the concrete microbars prepared with SJW by 16%–66% and 19%–67%, respectively.

Based on Figure 6, we can see, from the expansion rate, that adding 10%, 20%, and 30% fly ash reduced the expansion. However, when adding 30% fly ash, the expansion of SJW still exceeded 0.1%

at 90 d. For BFL8, when the alkali equivalent (equivalent Na_2Oeq) of the cement was 1.0%, adding 30% fly ash controlled the expansion within 0.1%.

3.3. The Effect of Fly Ash on Concrete Microbars in the Long-Term

The expansion results of the one-year concrete prisms test conducted on samples using 30% fly ash according to ASTM C1293 are shown in Figure 7, which shows that fly ash reduced the expansion for all tested aggregates.

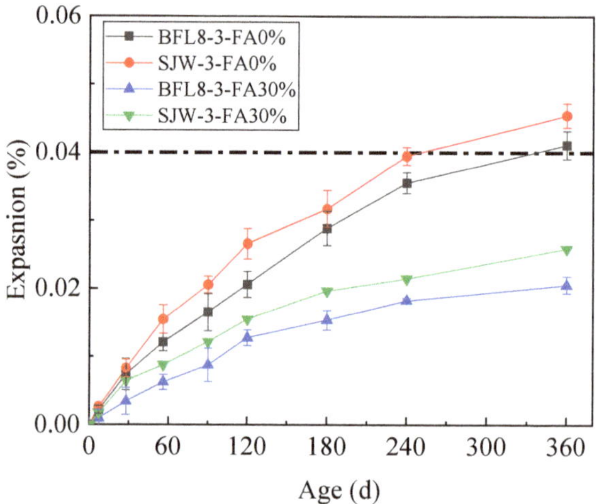

Figure 7. The expansion of the concrete prisms with fly ash.

Figure 7 shows that the expansion rates of BFL8 and SJW without fly ash were equal to 0.041% and 0.046%, respectively, at 360 days. In accordance with the expansion of the concrete prisms with 30% fly ash, using fly ash reduced the expansion due to the ACR, and thus the expansion rates of BFL8 and SJW decreased by 49% and 43%, respectively. The expansion of the concrete prisms prepared with 30% FA and made of BFL8 and SJW were 0.0206% and 0.0259%, respectively, at 360 d. The expansion rates of the concrete prisms were less than 0.040%. Therefore, adding 30% fly ash can effectively reduce the expansion rate due to the ACR.

3.4. The Mechanism of Fly Ash Inhibiting ACR

3.4.1. The Effect of Fly Ash on PH Value

The pH value of the cement paste pore solution cured in saturated $Ca(OH)_2$ at 80 °C for 28 days is shown in Figure 8, which indicates that fly ash can significantly reduce the pH value of the pore solution and delay the ACR reaction. The ACR reaction requires a pH value of 12, according to Tang and Den [22].

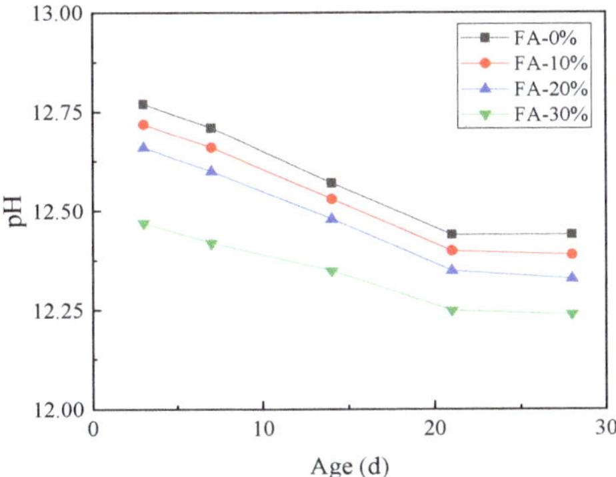

Figure 8. The influence of fly ash on pH in the cement paste pore solution (80 °C, Ca(OH)$_2$).

Adding 30% fly ash significantly decreased the pH value of the pore solution. Based on Figure 8, the pH value of the cement paste pore solution without fly ash was 12.45, and the pH of the cement paste pore solution with 30% fly ash was 12.25 at 21 days. Many researchers [23] have asserted that the concentration of OH$^-$ plays an important role in the ACR and that reducing the concentration of OH$^-$ can reduce the damage to the aggregates and inhibit the ACR.

The main reason why fly ash reduces the pH is because it plays the role of a physical diluent and reduces the alkali concentration of the pore solution [13,24]. Over the long-term, the fly ash reacts with Ca(OH)$_2$ to reduce the pH value.

3.4.2. The Effect of Fly Ash on the Ionic Migration and Pore Structure

Figure 9 shows the influence of fly ash on the concentrations of K$^+$ and Na$^+$ in the pore solution after curing in 1 mol/L NaOH solution at 80 °C. The concentrations of K$^+$ is shown in (A) and the concentrations of Na$^+$ is shown in (B).

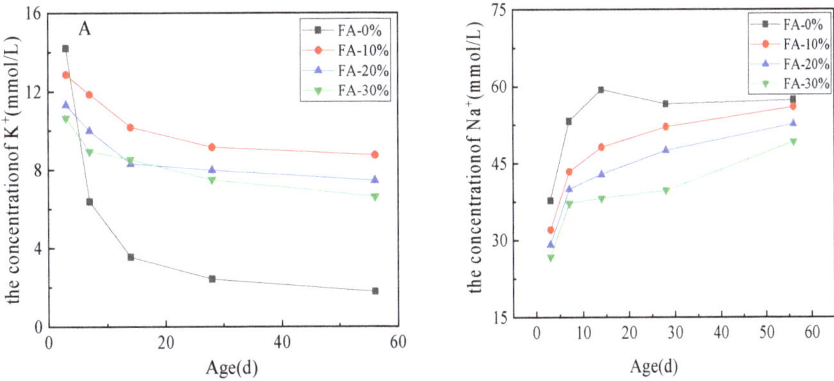

Figure 9. Effect of fly ash on the concentrations of K$^+$ and Na$^+$ in the pore solution (80 °C, 1 mol/L NaOH): (**A**) the concentrations of K$^+$ and (**B**) the concentrations of Na$^+$.

Figure 9 reveals that, as the amount of fly ash increased, the concentrations of K$^+$ and Na$^+$ in the pore solution decreased. Regardless of whether the sample was pulverized or not and mixed with fly ash, the concentration of K$^+$ ions in the pore solution gradually decreases and the concentration of Na$^+$ ions increases with the increasing curing time. Increasing the amount of fly ash slows the migration rates of Na$^+$ and K$^+$ to the cement paste in the curing solution. The ionic migration rate of the cement paste without fly ash was the fastest. The concentration of $[K^+ + Na^+]$ in the pore solution was close to 0.063 mol/L at 14 days, but the concentration of $[K^+ + Na^+]$ in the pore solution with 30% fly ash was only 0.055 mol/L at 56 days. There was still a significant difference between the cement paste with fly ash and cement paste without fly ash, which was not offset by the external alkali content. It can be seen that adding fly ash delays the ionic migration and exchange speed of the pore solution to the curing solution.

The pore structure is an important component of the concrete microstructure. Manmohan [25] divided the pores in concrete into four sizes: <4.5, 4.5–50, 50–100, and >100 nm. Mehta believed that only pores larger than 100 nm will affect the strength and permeability of concrete. We tested the effect of fly ash on the pore structure of the cement paste in the concrete microbars, in order to analyze the effect of fly ash on ionic migration. Figure 10 shows the pore volume and pore diameter distributions of the cement paste in the concrete microbars cured for 56 days in 1 mol/L NaOH solution at 80 °C. The porosity of cement paste in the concrete microbars is shown in (A) and the pore size distribution of cement paste in the concrete microbars is shown in (B).

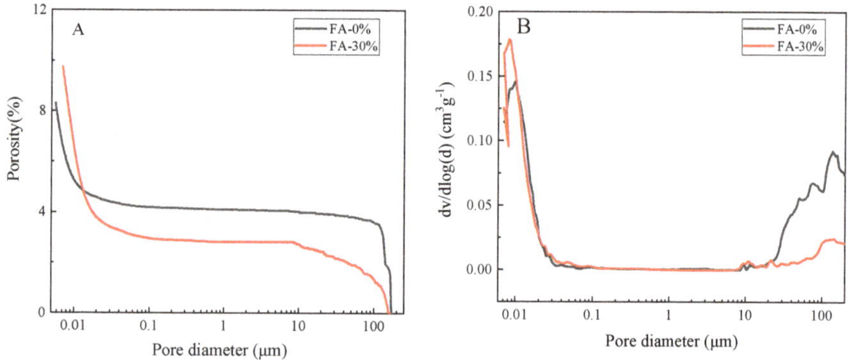

Figure 10. Pore structure of cement paste cured in 1 mol/L NaOH at 80 °C: (**A**) the porosity of cement paste in the concrete microbars and (**B**) the pore size distribution of cement paste in the concrete microbars.

Adding fly ash increased the porosity of the cement paste, reduced the harmful pores larger than 100 nm and increased the amount of micropores, effectively refining the pore structure. The porosity increased from 8% to 10% when using 30% fly ash. According to Manmohan [25], adding fly ash makes the structure of concrete microbars more compact. Fly ash not only reduces the content of the large pores, but also increases the content of small pores; thus, the pore structure of the cement paste becomes more compact, the permeability of the cement paste decreases, and the ionic migration speed becomes slower.

3.4.3. Polarizing Microscope and Stereomicroscope Analysis

The thin sections of the concrete microbars cured in NaOH solutions were observed by polarizing microscopy, as shown in Figure 11. From the figure, we can see the crack characteristics of the concrete prepared with the SJW and BFL8 aggregates in the NaOH solution. The cracks were created inside the rock and extended into the cement paste. We can see many different sizes of dolomite crystals distributed at the crack. Meanwhile, no ASR gel was found in the crack. Therefore, these cracks are expansion cracks, mainly due to the ACR.

Figure 11. Thin section of the concrete microbars cured in NaOH solution for 180 days: (**A**) concrete microbars prepared with SJW; and (**B**) concrete microbars prepared with BFL8.

The expansion cracks of the concrete microbars cured in 1 mol/L NaOH solution were examined by stereomicroscope analysis, as shown in Figure 12. From the figure, we can see that cracks appeared in the concrete microbars without fly ash. However, the concrete microbars mixed with 30% fly ash had only some fine cracks and even no cracking. This indicates that the addition of fly ash can significantly reduce the cracking in concrete microbars.

Figure 12. Expansion cracks of concrete microbars prepared with the dolomitic aggregate cured in NaOH solution for 56 days: (**A**) BFL8; (**B**) BFL8 with 30% fly ash; (**C**) SJW; and (**D**) SJW with 30% fly ash.

The above analysis shows that the dolomite in the rock had undergone a dolomitization reaction, and the cracks of the concrete microbars prepared with BFL8 and SJW were in dolomite-rich areas. Due to the fly ash refining the pore structure of the cement paste and reducing the migration rate of the alkali in the curing solution to the interior of the concrete microbars, as the content of fly ash increased, the concentration of K^+ and Na^+ and the pH value in the pore solution gradually decreased. This made the ACR in the rock slower, the cracks reduce, and the ACR expansion to be inhibited.

4. Conclusions

In this experiment, concrete microbars and concrete prisms made of dolomite aggregates, fly ash, and cement were used to systematically investigate the short- and long-term effects of fly ash on the ACR and to investigate the influence of fly ash on the expansion of dolomite aggregates. From the physical macro-measurements and microstructure analyses, the following main conclusions can be drawn:

Increasing the content of fly ash in concrete microbars and concrete prisms can considerably decrease the expansion rate due to the ACR. In comparison with adding 10% and 20% fly ash, concrete microbars prepared with 30% fly ash exhibited a greatly reduced expansion rate. According to the expansion rates of concrete microbars, when the alkali equivalent (equivalent Na_2Oeq) of the cement was 1.0%, for low-reactivity aggregates (such as BFL8), adding 30% fly ash could inhibit the expansion due to the ACR. However, when adding 30% fly ash to moderately reactive aggregates, such as SJW, although the expansion rate decreased, it did not decrease to less than 0.1% and cracking still occurred.

The polarizing microscope and stereomicroscope analysis results indicated that the cracks were expansion cracks, and the expansion was mainly due to the ACR. Adding 30% fly ash could significantly reduce the cracking in concrete microbars.

Through the analysis of the K^+ and Na^+ concentrations, the pH value, the ionic migration of the cement paste pore solution, and the pore structure, the main mechanism of fly ash inhibiting the ACR is that fly ash refines the pore structure of the cement paste, and the alkali migration rate in the curing solution to the interior of the concrete microbars is reduced. As the content of fly ash increases, the concentration of K^+ and Na^+ and the pH value in the pore solution gradually decrease. This causes the ACR in the rock to slow, the cracking to reduce, and the ACR expansion to be inhibited.

Author Contributions: Conceptualization, X.R. and M.D.; methodology, X.R.; software, X.R.; validation, X.R.,W.L., Z.M., and M.D.; formal analysis, X.R.; investigation, X.R.; resources, X.R., W.L. and Z.M.; data curation, X.R.; writing—Original draft preparation, X.R.; writing—review and editing, X.R.; visualization, X.R.; supervision, M.D.; project administration, M.D.; funding acquisition, M.D. All authors have read and agreed to the published version of the manuscript.

Funding: This work was funded by the National Key Research and Development Project (2016YFB0303601-2) and the Priority Academic Program Development of Jiangsu Higher Education Institutions (PAPD).

Acknowledgments: The authors gratefully acknowledge the assistance from Jun Wang, Pengcheng Yu and Huan Yuan from NJTECH, and the staff from State Key Laboratory of Materials-Oriented Chemical Engineering.

Conflicts of Interest: The authors declare no conflict of interest.

References

1. Stanton, T.E. Expansion of concrete through reaction between cement and aggregate. *Proc. Asce* **1940**, *6*, 1781–1811.
2. Gillott, J.E. Mechanism and kinetics of expansion in the alkali-carbonate rock reaction. *Can. J. Earth Sci.* **1964**, *1*, 121–145. [CrossRef]
3. Tang, M.; Liu, Z.; Han, S. Mechanism of alkali-carbonate reaction. In Proceedings of the 7th ICAAR in Concrete, Ottawa, ON, Canada, 18–22 August 1986; pp. 275–279.
4. Deng, M.; Mingshu, T. Mechanism of dedolomitization and expansion of dolomitic rocks. *Cem. Concr. Res.* **1993**, *23*, 1397–1408.
5. Tong, L.; Tang, M. Expansion mechanism of alkali-dolomite and alkali-magnesite reaction. *Cem. Concr. Compos.* **1999**, *21*, 361–373. [CrossRef]
6. Katayama, T.; Tagami, M.; Sarai, Y. Alkali-aggregate reaction under the influence of deicing salts in the Hokuriku district, Japan. *Mater. Charact.* **2004**, *53*, 105–122. [CrossRef]
7. Katayama, T. How to identify carbonate rock reactions in concrete. *Mater. Charact.* **2004**, *53*, 85–104. [CrossRef]
8. Katayama, T. The so-called alkali-carbonate reaction (ACR)—Its mineralogical and geochemical details, with special reference to ASR. *Cem. Concr. Res.* **2010**, *40*, 643–675. [CrossRef]
9. Chen, X.; Yang, B.; Mao, Z. The expansion cracks of dolomitic aggregates cured in TMAH solution caused by alkali–carbonate reaction. *Materials* **2019**, *12*, 1228. [CrossRef]
10. Chen, B.; Deng, M.; Lan, X. Behaviors of reactive silica and dolomite in tetramethyl ammonium hydroxide solutions. In Proceedings of the 15th international conference on Alkali-aggregate reactions in concrete, St Paul, Brazil, 3–7 July 2016; pp. 3–7.
11. Yuan, H.; Deng, M.; Chen, B. Expansion of dolomitic rocks in TMAH and NaOH solutions and its root causes. *Materials* **2020**, *13*, 308. [CrossRef] [PubMed]
12. Lindgård, J.; Andiç-Çakır, Ö.; Fernandes, I. Alkali–silica reactions (ASR): literature review on parameters influencing laboratory performance testing. *Cem. Concr. Res.* **2012**, *42*, 223–243. [CrossRef]
13. Kawabata, Y.; Yamada, K. The mechanism of limited inhibition by fly ash on expansion due to alkali–silica reaction at the pessimum proportion. *Cem. Concr. Res.* **2017**, *92*, 1–15. [CrossRef]
14. Thomas, M. The effect of supplementary cementing materials on alkali-silica reaction: A review. *Cem. Concr. Res.* **2011**, *41*, 1224–1231. [CrossRef]

15. Joshaghani, A. The effect of trass and fly ash in minimizing alkali–carbonate reaction in concrete. *Constr. Build. Mater.* **2017**, *150*, 583–590. [CrossRef]

16. Shehata, M.H.; Jagdat, S.; Rogers, C. Long-term effects of different cementing blends on alkali-carbonate reaction. *Aci Mater. J.* **2017**, *114*, 661–672. [CrossRef]

17. Min, D.; Mingshu, T. Measures to inhibit alkali-dolomite reaction. *Cem. Concr. Res.* **1993**, *23*, 1115–1120. [CrossRef]

18. AAR, R.R. Detection of potential alkali-reactivity of aggregates-The ultra-accelerated mortar-bar test. *Mater. Struct.* **2000**, *33*, 226.

19. Sommer, H.; Nixon, P.J.; Sims, I. AAR–5: Rapid preliminary screening test for carbonate aggregates. *Mater. Struct.* **2005**, *38*, 787–792. [CrossRef]

20. ASTM C1293–08: Standard Test Method for Determination of Length Change of Concrete Due To Alkali–Silica Reaction. Available online: https://books.google.com.hk/books?id=9ILMBQAAQBAJ&pg=PA8&lpg= PA8&dq=20.%09ASTM+C1293%E2%80%9308:+Standard+test+method+for+determination+of+length+ change+of+concrete+due+to+alkali%E2%80%93silica+reaction.&source=bl&ots=VxAaWxR3Oj&sig= ACfU3U2hRlpMOExB6rYk5bs8pXWL_UuNrA&hl=en&sa=X&redir_esc=y&hl=zh-CN&sourceid= cndr#v=onepage&q=20.%09ASTM%20C1293%E2%80%9308%3A%20Standard%20test%20method% 20for%20determination%20of%20length%20change%20of%20concrete%20due%20to%20alkali%E2%80% 93silica%20reaction.&f=false (accessed on 4 June 2020).

21. French, W.J. Concrete petrography: A review. *Q. J. Eng. Geol.* **1991**, *24*, 17–48. [CrossRef]

22. Deng, M.; Tang, M.S. Mechanism and prevention of alkali-dolomite reaction. *J. Nanjing Univ. Chem. Technol.* **1998**, *20*, 1–7.

23. Dunstan, E.R. The chemistry of alkali-aggregate reactions. *Cem. Concr. Aggreg.* **1981**, *3*, 101–104.

24. Xu, G.; Tian, Q.; Miao, J. Early-age hydration and mechanical properties of high-volume slag and fly ash concrete at different curing temperatures. *Constr. Build. Mater.* **2017**, *149*, 367–377. [CrossRef]

25. Mehta, P.K. Study on blended portland cements containing santirin earth. *Cem. Concr. Res.* **1981**, *11*, 507–518. [CrossRef]

 *crystals*

Article

Study on Visible Light Catalysis of Graphite Carbon Nitride-Silica Composite Material and Its Surface Treatment of Cement

Weiguang Zhong, Dan Wang, Congcong Jiang, Xiaolei Lu, Lina Zhang * and Xin Cheng *

Shandong Provincial Key Laboratory of Preparation and Measurement of Building Materials, University of Jinan, Jinan 250022, China; 20172120540@mail.ujn.edu.cn (W.Z.); mse_wangd@ujn.edu.cn (D.W.); mse_jiangcc@ujn.edu.cn (C.J.); mse_luxl@ujn.edu.cn (X.L.)

* Correspondence: mes_zhangln@ujn.edu.cn (L.Z.); mse_chengx@ujn.edu.cn (X.C.)

Received: 24 May 2020; Accepted: 5 June 2020; Published: 7 June 2020

Abstract: Cement-based composite is one of the essential building materials that has been widely used in infrastructure and facilities. During the service of cement-based materials, the performance of cement-based materials will be affected after the cement surface is exposed to pollutants. Not only can the surface of cement treated with a photocatalyst degrade pollutants, but it can also protect the cement-based materials from being destroyed. In this study, graphite carbon nitride-silica composite materials were synthesized by thermal polymerization using nanosilica and urea as raw materials. The effect of nanosilica content and specific surface area were investigated with the optimal condition attained to be 0.15 g and 300 m^2/g, respectively. An X-ray diffractometer, thermogravimetric analyzer, scanning electron microscope, a Brunauer–Emmett–Teller (BET) specific surface area analyzer and ultraviolet-visible spectrophotometer were utilized for the characterization of as-prepared graphite carbon nitride-silica composite materials. Subsequently, the surface of cement-based materials was treated with graphite carbon nitride-silica composite materials by the one-sided immersion and brushing methods for the study of photocatalytic performance. By comparing the degradation effect of Rhodamine B, it was found that the painting method is more suitable for the surface treatment of cement. In addition, through the reaction of calcium hydroxide and graphite carbon nitride-silica composite materials, it was found that the combination of graphite carbon nitride-silica composite materials and cement is through C-S-H gel.

Keywords: graphite carbon nitride; silica; visible light catalysis; cement

---

1. Introduction

In recent years, with the continuous improvement of the level of economic development and industrialization, pollution has become more and more serious, and the environment has been greatly damaged and even threatens human health [1,2]. In order to counter these problems, photocatalytic technology has attracted widespread attention as one of the most promising methods for controlling environmental pollution [3–6]. As the most widely used civil engineering material, cement-based composite is widely used in the construction sector in various countries and regions. It has been suggested to combine photocatalytic technology with building materials [7–10]. Based on building materials, photocatalytic catalysts are easily excited by the energy of sunlight and can degrade surrounding pollutants [11–13]. So far, many efforts have been made to combine TiO_2 photocatalysts with cement [14,15] to achieve photocatalytic degradation capabilities [16–18] and self-cleaning capabilities [19–23]. With the deepening of research, the emergence of some problems has also hindered the development of titanium dioxide photocatalytic cement [24,25]. First, TiO_2 with a wide band gap

(3.2 eV) cannot use visible light. Most of the energy of sunlight cannot be used by it, which causes energy to be wasted. Second, after the photocatalyst is coated on the cement surface, the binding between the catalyst and the cement is poor, which will cause the catalyst to fall off during use and thus reduce the photocatalytic efficiency. Therefore, a new type of visible light photocatalytic cement with a stable structure should be designed to solve the above problems.

As for the first problem, although there are a large number of methods reported to extend the spectral response range of photocatalysts [26–28], there have been few reports applied to cement materials so far. This may be due to their low activity and poor stability. However, the combination of cement and visible light photocatalysis technology will be an indispensable new requirement for the construction industry. Fortunately, graphite carbon nitride (g-C$_3$N$_4$) has been discovered as a stable visible light photocatalyst since 2008 [29]. Due to its narrow band gap (2.7 eV), g-C$_3$N$_4$ can make full use of visible light for photocatalytic water decomposition [30–32], organic matter degradation [33,34], and outdoor pollution control [35–37]. In contrast to traditional visible light photocatalysts, g-C$_3$N$_4$ is a polymer semiconductor similar to graphene and has good chemical and thermal stability [38,39].

Regarding the second problem, some scholars have prepared a SiO$_2$/g-C$_3$N$_4$ [40] composite material by heating a mixture of SiO$_2$ and melamine. The specific surface area of the obtained composite material increases, and the degree of aggregation of the graphite carbon nitride decreases. This can improve the catalyst's adsorption of pollutants so that the SiO$_2$/g-C$_3$N$_4$ composite has higher activity in the process of the photocatalytic degradation of pollutants. In addition, highly reactive nanosilica can react with cement to produce C-S-H gel. The functional layer and the cement matrix are combined by C-S-H gel [41]. This provides a way to counter the problem of bonding.

In this paper, by controlling the specific surface area and the additional amount of nanosilica, the optimal preparation conditions of graphite carbon nitride-silica composite materials are discussed. The effect of nanosilica on the modification of graphite carbon nitride is explored by XRD, TGA SEM, UV–Vis, and Brunauer–Emmett–Teller (BET) methods. Subsequently, the prepared graphite carbon nitride-silica composite material is used to treat the cement surface, and its photocatalytic ability is studied by degrading the dye under visible light. At the same time, the reaction mechanism of graphite carbon nitride-silica composite material and calcium hydroxide is explored.

2. Materials and Methods

2.1. Materials

Rhodamine B (RhB) and urea were purchased from Sinopharm Chemical Reagent Co. Ltd. (Shanghai, China) without any purification. RhB was prepared as a 10 mg/l solution for use. Nanosilica was purchased from Aladdin Reagent Co. Ltd. (Shanghai, China). Its specific surface areas are 200 and 300 m^2/g, respectively. P.W 525 white cement was purchased from Shandong Shanshui Cement Co. Ltd. (Jinan, Shandong, China). Its density is 3.2 g/cm^3. Distilled water obtained from a water purification system (Direct-Q® 3.5.8, Millipore Co. Ltd., Burlington, MA, USA).

2.2. Preparation of Graphite Carbon Nitride-Silica Composite Materials

First, 50 mL deionized water was added to a 100 mL crucible. Nanosilica with a specific surface area of 200 m^2/g was then added to the crucible. The amount of nanosilica added was controlled to 0.05 g, 0.10 g, 0.15 g, and 0.20 g. The nanosilica solution was sonicated for 30 min by using an ultrasonic cell grinder. The purpose was to uniformly disperse the nanosilica in deionized water, and then 12 g urea was added to the crucible. After stirring for 30 s with a glass rod, the crucible was placed in an ultrasonic cleaning machine for 30 min for ultrasonic treatment, with the purpose to completely dissolve the urea. After the treatment, the crucible was dried in an oven at 70 °C. After drying, the crucible was taken out; the crucible mouth was covered with aluminum foil paper, and the crucible lid was covered. The crucible was placed in a muffle furnace and heated to 550 °C in an air atmosphere. The temperature increase rate was 5 °C/min, and the holding time was 2 h.

The obtained graphite carbon nitride-silica composite material (g-C_3N_4-SiO_2) was taken out from the crucible after natural cooling, and it could then be used after grinding. By adjusting the specific surface area of SiO_2 to 300 m^2/g, several other samples were prepared by the same method. It was named CS005-200, CS011-200, CS015-200, CS020-200, CS005-300, CS010-300 CS015-300, CS020-300 according to the amount and specific surface area of nanosilica.

2.3. Characterization of g-C_3N_4-SiO_2

The crystal structure of the sample was determined by an X-ray diffractometer (XRD, D8 Advance, Bruker Co. Ltd., Karlsruhe, Ban-Württemberg, Germany). A thermal gravimetric analyzer (TGA/DSC 1, Mettler, Switzerland) was used to determine the sample composition. The field emission scanning electron microscope (QUANTA 250 FEG, FEI Co. Ltd., Hillsboro, OR, USA) was used to observe the micromorphology of the samples. Its acceleration voltage is 20 kV. The band gap was characterized by a UV–Vis DRS spectrum by an ultraviolet-visible spectrophotometer (U-4100 Hitachi Co. Ltd., Tokyo, Japan). The photocatalytic activity of g-C_3N_4-SiO_2 was evaluated by the degradation of RhB (10 mg/l) under visible light in a photo reactor (Beijing Princes Co. Ltd., Beijing, China). The light source is a xenon lamp with a power of 350 W. The absorbance of RhB was measured by an ultraviolet-visible spectrophotometer (U-4100, Hitachi Co. Ltd., Tokyo, Japan). The degradation rate was calculated by dividing the absorbance of the sample by its original absorbance. The specific surface area was measured by the Brunauer–Emmett–Teller (BET) method using a specific surface area analysis tester (MFA-140 Beijing Peaudi Co. Ltd., Beijing, China).

2.4. Surface Treatment of Cement with g-C_3N_4-SiO_2

Cement paste produced with white cement was made into a cube of 20 mm × 20 mm × 20 mm with a water–cement ratio of 0.35. The cement sample was placed in the curing room for 7 days. The g-C_3N_4-SiO_2 was mixed with water to make a 1 mg/mL suspension. The photocatalyst was attached to the surface of cement by the one-sided immersion method and the brushing method. For the one-sided immersion method, the flank of cement was covered with adhesive tape, and the top surface was surrounded. The suspension was sucked by using a dropper, and it was evenly dropped to the top surface. For the brushing method, the dispersion liquid was dipped by using a brush and then repeatedly brushed onto the cement surface. The treated cement samples were placed in a dark curing cabinet for 7 days. The temperature was 25 °C. The humidity was 50%.

2.5. Evaluation of Visible Light Catalytic Performance of Photocatalytic Cement

The cement sample was taken out of the curing box. The RhB solution (10 mg/L) was sprayed evenly onto the treated surface. The sprayed cement was put in the dark for curing for 24 h. After the RhB solution on the surface of the sample was dried, it was taken out of the curing box. It was placed in the photocatalytic reactor and the test surface was aimed at the light source. Under the condition of visible light irradiation, the picture was taken every 30 min, and its color change was compared.

2.6. Exploration of the Binding Mechanism of Graphite Carbon Nitride-Silicon Dioxide Composite Materials and Cement

We added 60 ml of a saturated calcium hydroxide solution to the plastic bottle. Then, 0.15 g of CS015-300 was added to the solution. After stirring evenly, the pH value of the solution was measured with a pH test paper. The plastic bottle was sealed, and it was stirred at 1000 rpm on the magnetic stirrer for 7 days. After stirring, the resulting sample was taken out, and the pH of the solution was measured again. After the measurement was completed, the sample was dried in a vacuum drying oven, and then it was ground and tested.

3. Results and Discussion

3.1. Crystal Structure

In order to analyze the crystal structure of the synthesized of g-C_3N_4-SiO_2, we characterized it with an X-ray diffractometer. Figure 1a is the XRD pattern of g-C_3N_4-SiO_2 synthesized with SiO_2 with a specific surface area of 200 m^2/g. There is no obvious diffraction peak in the XRD pattern of SiO_2, but there is a steamed bread around $2\theta = 23°$. The reason is that gas-phase nano-SiO_2 is amorphous [42]. This is also the reason why there are only two diffraction peaks of $2\theta = 12.9°$ and $2\theta = 27.5°$ in the XRD pattern of the g-C_3N_4-SiO_2. Here, $2\theta = 12.9°$ and $2\theta = 27.5°$ correspond to the (100) and (002) crystal planes of g-C_3N_4, respectively [43]. After the addition of SiO_2, the intensity of the diffraction peak at $2\theta = 12.9°$ is weaker than that of the single g-C_3N_4. With the increase of the amount of SiO_2 added, the intensity of the diffraction peak showed a trend of gradually decreasing. The change law of the diffraction peak intensity at $2\theta = 27.5°$ is consistent with the change law of the diffraction peak intensity at $2\theta = 12.9°$. This may be related to the decrease of the proportion of g-C_3N_4 when the amount of SiO_2 in the sample is increased. Figure 1b is the XRD pattern of g-C_3N_4-SiO_2 synthesized with SiO_2 with a specific surface area of 300 m^2/g. An analysis of the changes in the diffraction peaks of the two graphs reveals that they are basically the same as those shown in Figure 1a. Their rule is that there is no change in the position of the diffraction peak, and the intensity of the same diffraction peak decreases as the amount of SiO_2 added increases. This can also be attributed to the reduction of graphite carbon nitride content in the sample. From a comprehensive analysis of Figure 1a,b, it can be concluded that the change in the specific surface area of SiO_2 will not affect the crystal structure of the g-C_3N_4-SiO_2.

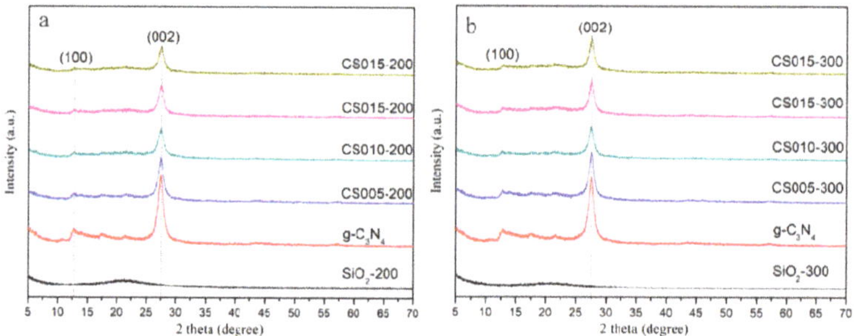

Figure 1. The XRD patterns of SiO_2-200, g-C_3N_4, CS005-200, CS010-200, CS015-200, CS020-200 (**a**) and SiO_2-300, g-C_3N_4, CS005-300, CS010-300, CS015-300, CS020-300 (**b**).

3.2. Composition

Figure 2 shows the TG curve of SiO_2, g-C_3N_4 and g-C_3N_4-SiO_2, CS005-200, CS010-200, CS015-200, CS020-200, CS005-300, CS010-300, CS015-300, CS020-300. Below 580 °C, the quality of g-C_3N_4 is basically unchanged. This shows that g-C_3N_4 has stable chemical properties below 580 °C. The reason is that the internal structure is an aromatic ring conjugated system connected by a covalent bond of carbon and nitrogen. On the other hand, g-C_3N_4 has a layered structure similar to graphite. The van der Waals force between layers is relatively strong [44]. From the beginning of 580 °C, as the temperature is increased, g-C_3N_4 is decomposed. The g-C_3N_4 is completely decomposed at 750 °C. At this time, the mass of g-C_3N_4 is 0. This corresponds to the position of the inflection point in the curve. SiO_2 basically has no mass loss below 900 °C. It has excellent thermochemical stability below 900 °C [45]. According to the characteristics of g-C_3N_4 and SiO_2 in the TG curve, we can determine the proportion of SiO_2 in different samples according to the value of the inflection point in the curve. In CS005-200, the content of SiO_2 is 16.48%. In CS010-200, the content of SiO_2 is 29.66%. In CS015-200, the content of SiO_2 is 31.24%. In CS020-200, the content of SiO_2 is 35.62%. In CS005-300, the content of

SiO$_2$ is 12.40%. In CS010-300, the content of SiO$_2$ is 28.41%. In CS015-300, the content of SiO$_2$ is 36.26%. In CS020-300, the content of SiO$_2$ is 41.28%.

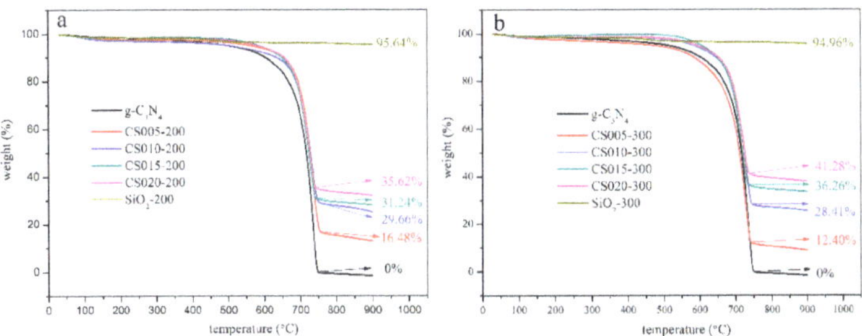

Figure 2. The TGA analysis of SiO$_2$-200, g-C$_3$N$_4$, CS005-200, CS010-200, CS015-200, CS020-200 (**a**) and SiO$_2$-300, g-C$_3$N$_4$, CS005-300, CS010-300, CS015-300, CS020-300 (**b**).

3.3. Surface Morphology and Structure

As shown in Figure 3, the structure of CS005-200 is a stacked layer. The spacing between the slices is very obvious; there are holes in the middle of the block. At the same time, there are a few spherical SiO$_2$ particles in the middle of the block, which are attached to both sides of the part of the plate. Obviously, more spherical SiO$_2$ particles were observed in CS010-200 than in CS005-200. The degree of stacking between the sheets becomes weak. The form is developed into a three-dimensional structure. There are more g-C$_3$N$_4$ flakes in the sample than spherical SiO$_2$ particles, which leads to uneven distribution. The spherical SiO$_2$ particles in CS015-200 are evenly embedded in the holes of the g-C$_3$N$_4$ sheet. In CS020-300, spherical SiO$_2$ increased significantly, while g-C$_3$N$_4$ flakes decreased. The distribution of SiO$_2$ and g-C$_3$N$_4$ is also uneven. As the SiO$_2$ content increases, more SiO$_2$ particles are introduced into the pores of the g-C$_3$N$_4$ flake. When the added amount reached 0.15 g, the SiO$_2$ particles evenly entered the pores of the g-C$_3$N$_4$. At this time, the modification effect is the best. The amount of SiO$_2$ continued to increase, and SiO$_2$ was distributed on the surface. Moreover, the porosity in the sample is also reduced, which leads to a poor modification effect. The SEM changes of CS005-300, CS010-300, CS015-300, and CS020-300 are similar to those of the CS200 series. It should be noted that the holes in the CS300 series samples are smaller than those in the CS200 series. The increase in porosity will increase the specific surface area, thereby improving the photocatalytic activity. Therefore, when the specific surface area of SiO$_2$ is 300 m^2/g SiO$_2$ and the addition amount is 0.15 g, the obtained g-C$_3$N$_4$-SiO$_2$ has the best performance. The results of the EDS analysis are shown in Figure 4. CS015-300 contains four elements of C, N, Si, and O. Both C and N come from g-C$_3$N$_4$, while Si and O come from SiO$_2$. This indicates that g-C$_3$N$_4$-SiO$_2$ has been successfully synthesized.

Figure 3. SEM images of CS005-200, CS010-200, CS015-200, CS020-200, CS005-300, CS010-300, CS015-300, CS020-300.

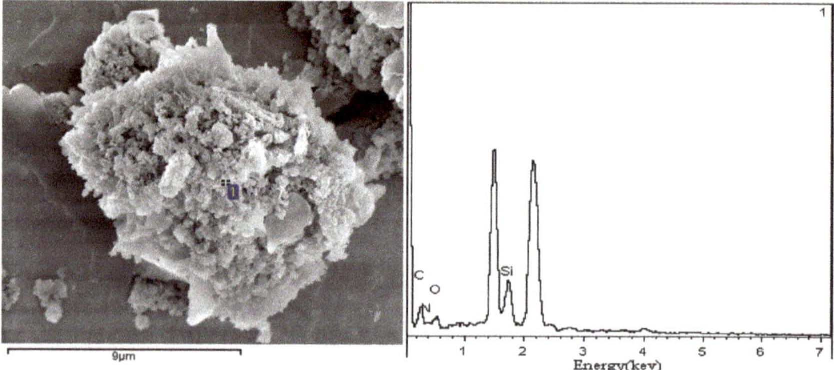

Figure 4. The EDS analysis of CS015-300.

3.4. Specific Surface Area and Porosity

Figure 5 is the nitrogen adsorption and desorption isotherms of g-C_3N_4, CS015-200, and CS015-300. The specific surface area, the total pore volume and the average pore radius measured by the N_2 adsorption-desorption experimental data and the BET model are listed in Table 1. The adsorption-desorption curves of g-C_3N_4 and CS015-300 indicate a type Π isotherm, which is the characteristic of microporous materials. It can be seen from Table 1 that after the addition of nano-SiO_2, the specific surface area of the sample became larger, the total pore volume became smaller, and the average pore radius became smaller. The specific surface area of CS015-200 was increased by 61.68%, the total pore volume was expanded by 56.13%, and the average pore radius was reduced by 3.41%. The specific surface area of CS015-300 was increased by 126.35%, the total pore volume was expanded by 15.70%, and the average pore radius was reduced by 52.99%. The reason is that more pores were introduced with the addition of nano-SiO_2, which caused the pore volume to be reduced and the pore size to become smaller. This caused the specific surface area to be increased. Although it was concluded from the previous analysis that the addition of SiO_2 has little effect on the band structure of the sample, the specific surface area has greatly improved. As the specific surface area is increased, more active sites are provided [46]. This is conducive to the improvement of catalytic capacity.

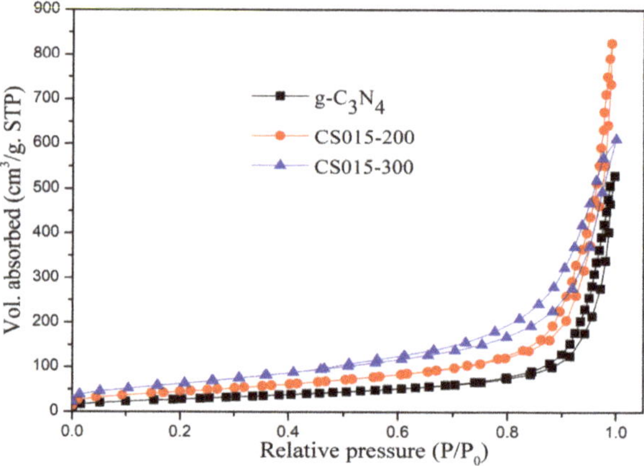

Figure 5. N_2 adsorption-desorption isotherms of g-C_3N_4, CS015-200, and CS015-300.

Table 1. The specific surface area, total pore volume, and average hole radius of g-C_3N_4, CS015-200 and CS015-300.

Sample	Specific Surface Area	Total Pore Volume	Average Hole Radius
g-C_3N_4	105.45 m^2/g	0.81 cm^3/g	15.53 nm
CS015-200	170.49 m^2/g	1.27 cm^3/g	15.00 nm
CS015-300	238.68 m^2/g	0.94 cm^3/g	7.30 nm

3.5. Band Gap and Absorption Edge

As can be seen from Figure 6a, the UV–Vis spectral curve of CS005-200 and of CS010-200 are similar to that of g-C_3N_4. The absorption edge does not move. The absorption edge of CS015-200 has a slight red shift. The UV–Vis spectrum of CS020-200 is different from that of the previous materials; its absorption edge has a slight blue shift. Their change rule is that the UV–Vis spectrum of the sample does not change when the amount of SiO_2 is low. As the amount of SiO_2 is increased, the absorption edge of the sample is red-shifted, but the degree of shift is not large. As the amount of SiO_2 is continuously increased, the absorption edge of the sample has a blue shift. This shows that the low additional amount of SiO_2 will not affect the visible light absorption range. The addition of SiO_2 will expand the visible light absorption range of the sample under the appropriate addition amount. However, when the blending amount is too large, the negative effect will appear. From Figure 6b, we can find that the width of the band gap of CS005-200 and CS010-200 is the same as that of g-C_3N_4, both of which are 2.7 eV. The width of the forbidden band of CS015-200 was reduced to 2.65 eV, and the absorption edge was increased to 471 nm. The band gap of CS020-200 was expanded to 2.8 eV, and the absorption edge was reduced to 468 nm. The variation law of the band gap of each sample is the same as that of its corresponding ultraviolet-visible diffuse reflection spectrum.

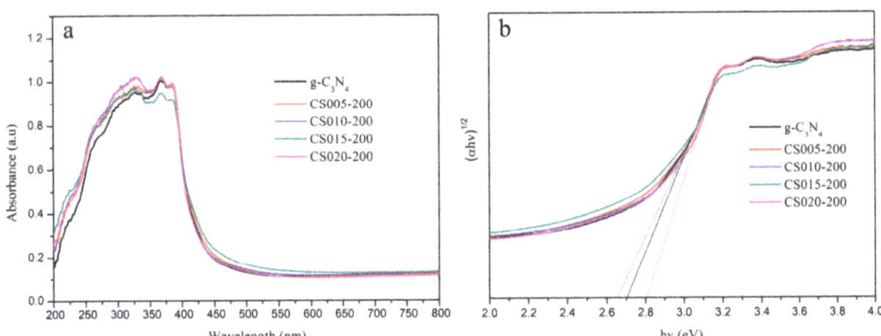

Figure 6. UV–Vis spectra (**a**) and band gap (**b**) of g-C_3N_4, CS005-200, CS010-200, CS015-200, CS020-200.

The UV–Vis spectrum of the g-C_3N_4-SiO_2 prepared with SiO_2 with a specific surface area of 300 m^2/g is shown in Figure 7a. The absorption edge of the sample changed. The absorption edge of CS005-300 and CS020-300 has a slight blue shift compared to g-C_3N_4. The movement of CS005-300 was slightly larger than that of CS020-300. The curve of CS010-300 coincides with the curve of g-C_3N_4, and the absorption edge did not change. A slight red shift occurred in the absorption edge of CS015-300. The difference with the graphite carbon nitride-silica (S_{BET} = 200 m^2/g) composite material is that the absorption edge of the sample had a blue shift at lower dosing levels and too much doping. The corresponding absorption of visible light was also reduced. From Figure 7b, it can be found that the band gap of CS005-300 expanded to 2.8 eV, and the absorption edge was reduced to 442 nm. The band gaps of CS010-300 and CS015-300 were reduced to 2.68 eV, and the absorption edge expanded to 462 nm. The band gap of CS020-300 expanded to 2.75 eV, and the absorption edge was reduced to 451 nm. The band gap of g-C_3N_4 is still 2.7 eV. The absorption edge is 460 nm. Because the added SiO_2 is amorphous, the changes in the band gap and the absorption edge of g-C_3N_4-SiO_2 were very slight. The change in band gap is due to the quantum size effect of the smaller size g-C$_3$N$_4$ generated on the surface of nano-SiO$_2$ [47].

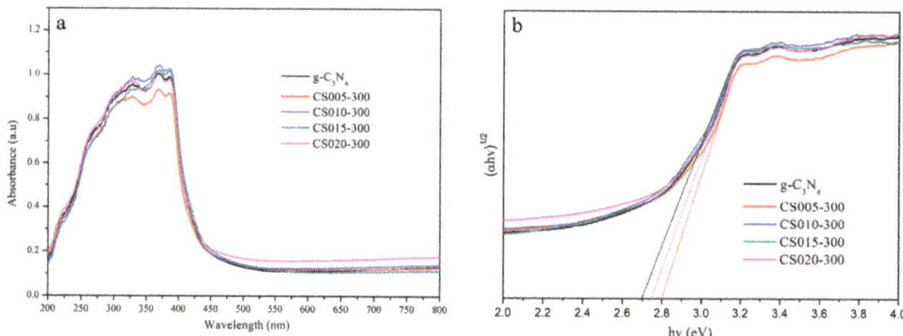

Figure 7. UV–Vis spectra (**a**) and band gap (**b**) of g-C$_3$N$_4$, CS005-300, CS010-300, CS015-300, CS020-300.

3.6. Photocatalytic Evaluation

As shown in Figure 8a, RhB is not degraded without adding a catalyst. After 1 h of dark treatment, g-C$_3$N$_4$ has the same dye adsorption rate as the other four samples, which is about 27%. With the extension of visible light exposure time, the absorbance of the dye is gradually reduced, and the dye is continuously degraded. The minimum degradation time of CS015-200 is 55 min. Secondly, the degradation time of CS005-200 is 60 min. Thirdly, the degradation time of CS010-200 is 70 min. Fourthly, the degradation time of g-C$_3$N$_4$ is 75 min. The maximum degradation time of CS020-200 is 90 min. The catalytic performance from strong to weak is ranked as CS015-200, CS005-200, CS010-200, g-C$_3$N$_4$, CS020-200. The reason for the reduced efficiency of CS020-200 may be related to the reduced visible light absorption range caused by the reduction of the absorption edge. Among all the samples, the improvement of the CS015-200 is the most obvious, and the performance has improved by 26.67% compared with the g-C$_3$N$_4$.

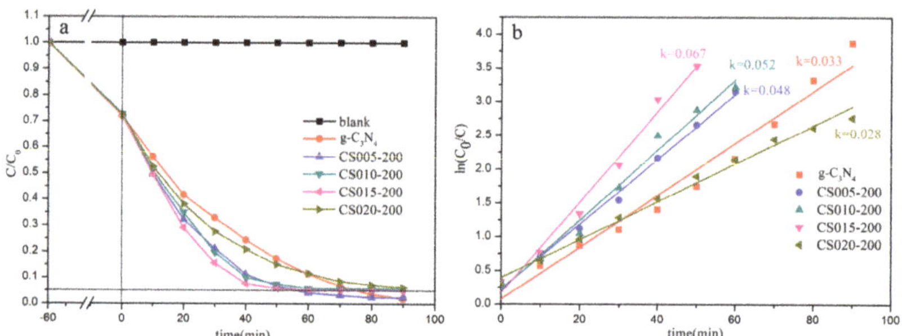

Figure 8. Photocatalytic activities (**a**) and their kinetic constants (**b**) of g-C$_3$N$_4$, CS005-200, CS010-200, CS015-200, CS020-200.

From Figure 8b, it can be concluded that the largest k value of CS015-200 is 0.067. Secondly, the k value of CS010-200 is 0.052. Thirdly, the k value of CS005-200 is 0.048. Fourthly, the k value of g-C$_3$N$_4$ is 0.033. The minimum k value of CS020-200 is 0.025. After silica is added, the k value of each sample is changed to varying degrees. CS015-200 is 103.03% higher than g-C$_3$N$_4$. CS010-200 is 57.57% higher than g-C$_3$N$_4$. CS005-200 is 45.45% higher than g-C$_3$N$_4$, and CS020-200 is 24.24% lower than g-C$_3$N$_4$. The performance of CS015-200 is the best among all the samples in the picture.

As shown in Figure 9a, after 1 h of dark treatment, different samples have different adsorption rates for dyes. The adsorption rate of composite materials is higher than that of the single g-C$_3$N$_4$, and the adsorption rate of g-C$_3$N$_4$ is about 27%. The adsorption rate of CS010-300 and CS020-300 is about 33%. The adsorption rate of CS005-300 and CS015-300 is about 39%. The minimum degradation time of CS015-300 is 50 min. Secondly, the degradation time of CS005-300 and CS010-300 is 60 min. Thirdly, the degradation time of CS0020-300 is 70 min. The maximum degradation time of g-C$_3$N$_4$ is 75 min. The catalytic performance from strong to weak is ranked as CS015-300, CS005-300 (CS010-300), CS0020-300, g-C$_3$N$_4$. Compared with g-C$_3$N$_4$, the visible light catalytic ability of each sample has improved. This shows that the visible light catalytic ability of the sample is increased with the addition of SiO$_2$ with a specific surface area of 300 m^2/g.

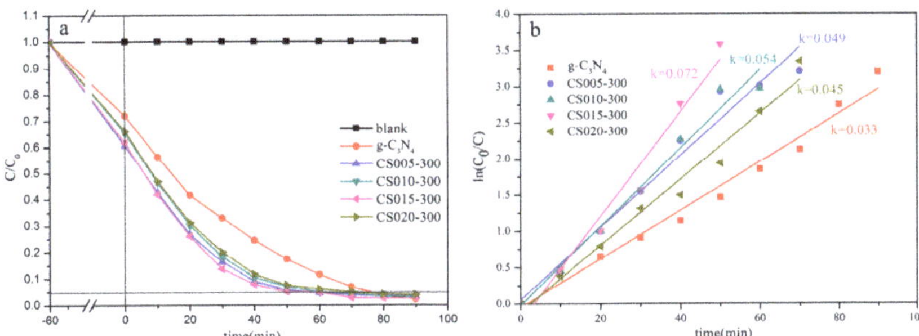

Figure 9. Photocatalytic activities (**a**) and their kinetic constants (**b**) of g-C$_3$N$_4$, CS005-300, CS010-300, CS015-300, CS020-300.

From Figure 9b, the maximum k value of CS015-300 is 0.072. Secondly, the k value of CS010-300 is 0.054. Thirdly, the k value of CS010-300 is 0.049. Fourthly, the k value of CS020-300 is 0.045. The minimum k value of g-C$_3$N$_4$ is 0.033. The difference from Figure 9 is that after adding nanosilica with a specific surface area of 300 m^2/g, the k value of all samples has improved. CS015-300 is 118.18% higher than g-C$_3$N$_4$. CS010-200 is 63.63% higher than g-C$_3$N$_4$. CS005-200 is 48.48% higher than g-C$_3$N$_4$. CS020-200 is 36.36% higher than g-C$_3$N$_4$. The performance of CS015-300 is the best among all the samples in the picture.

By comparing the degradation rate constants of the two types of samples, we find that the g-C$_3$N$_4$-SiO$_2$ prepared using SiO$_2$ with a specific surface area of 300 m^2/g has the best visible light catalytic performance. When the same amount is added, the degradation rate constant is always higher than that of g-C$_3$N$_4$-SiO$_2$ prepared with SiO$_2$ with a specific surface area of 200 m^2/g. In addition, the degradation rate constant of heterogeneous graphite carbon nitride-silica prepared with SiO$_2$ with a specific surface area of 300 m^2/g is higher than that of g-C$_3$N$_4$. The phenomenon of reduced catalytic performance did not occur.

3.7. Surface Treatment Evaluation

As can be seen from Figure 10, the catalyst distribution on the surface of the cement material treated by the one-sided immersion method is uneven, and the thickness of the coating is different. The color in the middle of the sample is obviously lighter than the color at the edge of the sample. This shows that there are more catalysts on the cement edge than in the middle. Cracking and shedding of the coating appear on the edge of the cement block. The reason is that the solution was dripping onto the cement surface, and the liquid level is affected by the surface tension; it appears to be high in the surroundings and low in the middle [48]. As the water evaporated, the water in the middle disappeared firstly and then gradually spread to the surroundings. This caused the surrounding coating to be too thick, which caused cracking and shedding. The color distribution of the surface of the cement material treated was relatively uniform by the brushing method, which indicates that the distribution of the catalyst was relatively uniform. There is no cracking and shedding of the catalyst on the cement surface, which indicates that the catalyst did not appear on the cement surface. The reason is that the brush dipped in less solution during the application process, and when applied to the cement surface, the surface tension is less affected. At the same time, the catalyst is further uniformly dispersed in the process of the repeated three times of brushing, thereby avoiding cracking and shedding caused by the aggregation of the catalyst.

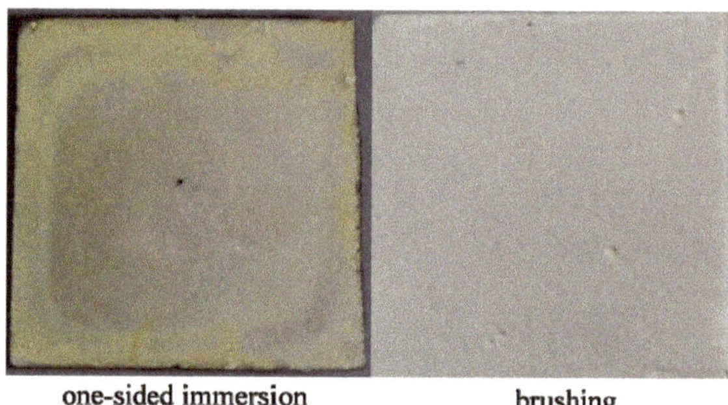

Figure 10. Surface treatment of cement by the one-sided immersion method and brushing method.

3.8. Performance Evaluation of Photocatalytic Cement

Figure 11 is the color change of RhB after the photocatalytic cement prepared by the one-sided immersion method was exposed to visible light. As shown in Figure 11, comparing the photos of the cement under visible light irradiation, it was found that with the increase of light time, the color of the surface without surface treatment is basically unchanged, but the color of all surface-treated samples was lightened to varying degrees. This shows that the cement had a self-cleaning function after surface treatment. The reason is that the catalyst on the cement surface captures the pollutants, and the dye was degraded under the irradiation of visible light. However, the self-cleaning efficiency of cement materials treated with different catalysts varies. After 60 min of irradiation, the degree of discoloration is ranked as CS015-300, CS010-300, CS005-300, CS020-300, g-C_3N_4 in order from large to small. This is consistent with the rule of g-C_3N_4-SiO_2 degrading RhB in solution. This shows that the performance of the catalyst did not change after being applied to the cement surface.

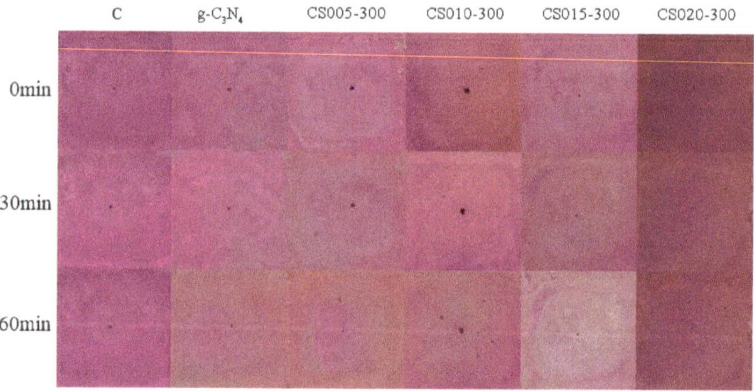

Figure 11. Photocatalytic performance of surface treatment cement with the one-sided immersion method.

Figure 12 is the color change of RhB after the photocatalytic cement prepared by the brushing method was exposed to visible light. After 60 min of irradiation, the degree of discoloration was ranked as CS015-300, CS010-300, CS005-300, CS020-300, g-C_3N_4 in order from large to small. The discoloration of CS015-300 is most obvious. However, since the catalyst dispersion on the cement surface treated by the brushing method was more uniform, the color change was more obvious than that of the one-sided immersion method. The performance of the photocatalytic cement obtained by the brushing method was more excellent.

Figure 12. Photocatalytic performance of surface treatment cement by the brushing method.

3.9. Exploration of Binding Mechanism

In order to explore the binding mechanism of g-C_3N_4-SiO_2 and cement, we used saturated calcium hydroxide solution to simulate the alkaline environment of cement, and then an appropriate amount of CS015-300 was added. Under sealed conditions, the two substances are reacted. The resulting product was analyzed.

As can be seen from Figure 13, after adding CS015-300 to the saturated calcium hydroxide solution, the pH value of the test mixed solution is 13. After 7 days of reaction, the pH of the solution was reduced to 9. This indicates that the content of OH- in the solution was reduced, and a part of calcium hydroxide was reacted because SiO_2 can react with calcium hydroxide by pozzolanic reaction [49]. According to existing research, it has been found that the product is C-S-H gel after the reaction [41,50]. We think that the nanosilica in CS015-300 reacted with the calcium hydroxide in the solution to form C-S-H, and the calcium hydroxide was consumed.

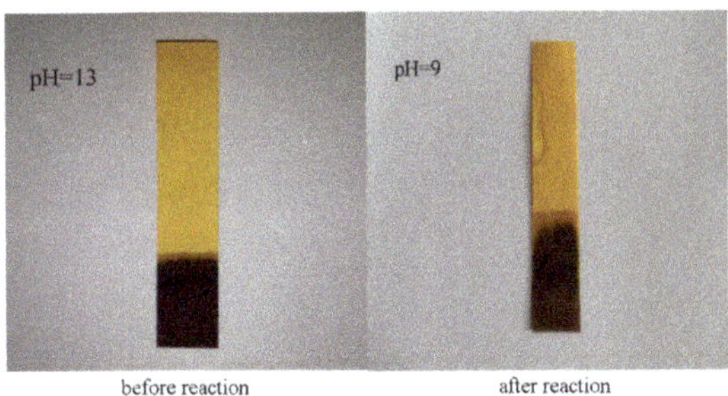

Figure 13. The pH of solution before and after reaction.

As can be seen from Figure 14, comparing the SEM pictures of the samples before and after the reaction, it was found that the microscopic morphology of the sample changed significantly after the reaction with calcium hydroxide. The microstructure of CS015-300 changed from the previous nanosilica embedded in the pores of g-C_3N_4 to the accumulation of irregular lumps. After further zooming, the rod was found. This point was analyzed by EDS, and the results are shown in Figure 15. It can be observed from Figure 15 that the main chemical composition of the rod is C, N, Si, O, and Ca. By analyzing the distribution of the four elements C, N, Si, and Ca in the pink area, it was found that the distribution of the four elements is very uniform and no aggregation occurred. This further illustrates that CS015-300 reacted with calcium hydroxide rather than simply piled together. This can further explain that CS015-300 reacted with calcium hydroxide to form a C-S-H gel [50]. Through the simulation experiments of CS015-300 and calcium hydroxide, we can draw the conclusion that the calcium hydroxide produced by cement hydration reacts with the nanosilica in the composite material to form C-S-H gel. The g-C$_3$N$_4$-SiO$_2$ and cement material are combined by C-S-H gel.

the SEM of CS015-300 before reaction

the SEM of CS015-300 after reaction

the SEM of CS015-300 after reaction

Figure 14. The SEM of CS015-300 before and after reaction.

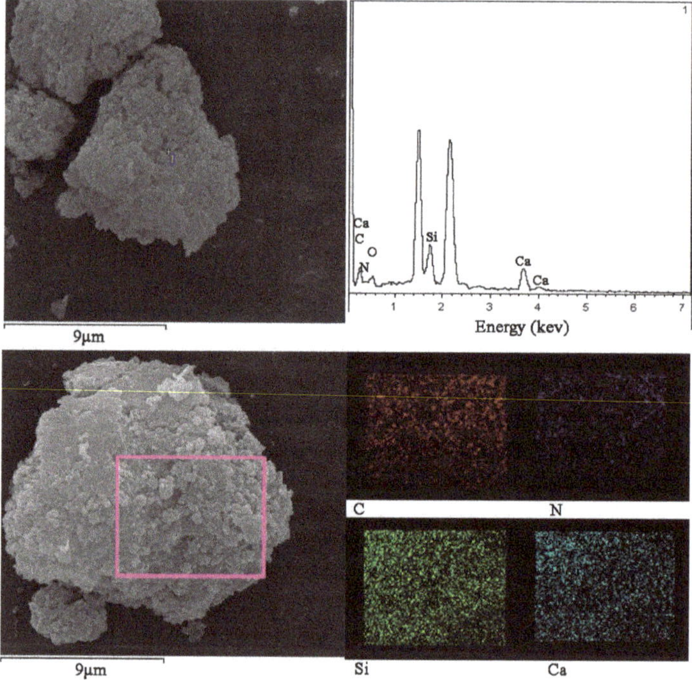

Figure 15. The EDS of CS015-300 after reaction.

4. Conclusions

In summary, the g-C$_3$N$_4$-SiO$_2$ was synthesized by mixing and heating nanosilica and urea. By controlling the amount and the specific surface area of nanosilica, the best synthesis conditions were selected. In this way, the band gap of the graphite carbon nitride was changed, and the specific surface area was increased. The visible light catalytic performance of the material has improved.

After the graphite carbon nitride-silicon dioxide composite material was treated by brushing and by one-sided immersion, the cement obtained the photocatalytic function. By comparing the photocatalytic efficiency, the brushing method was more suitable for the surface treatment of cement than the one-sided immersion method. The combination mechanism of calcium hydroxide and g-C_3N_4-SiO_2 was explored. It was found that g-C_3N_4-SiO_2 and cement material were combined by C-S-H gel.

Author Contributions: Conceptualization: W.Z. and X.L.; methodology: all authors; validation: W.Z., D.W., and L.Z.; formal analysis: all authors; investigation: W.Z., L.Z., and C.J.; resources: all authors; writing—original draft preparation: W.Z.; writing—review and editing: all authors; visualization: all authors; supervision: all authors; project administration: L.Z.; funding acquisition: L.Z. and X.C. All authors have read and agreed to the published version of the manuscript.

Funding: This research was funded by the Program for Taishan Scholars Program, the Case-by-Case Project for Top Outstanding Talents of Jinan, the Distinguished Taishan Scholars in Climbing Plan, the Science and Technology Innovation Support Plan for Young Researchers in Institutes of Higher Education in Shandong (2019KJA017), the National Natural Science Foundation of China (Grant No. 51872121, 51632003, and 51902129), the National Key Research and Development Program of China (Grant No. 2016YFB0303505) and the 111 Project of International Corporation on Advanced Cement-based Materials (No.D17001).

Acknowledgments: The authors wish to gratefully thank Shandong Provincial Key Laboratory of Preparation and Measurement of Building Materials, University of Jinan for their support of this work.

Conflicts of Interest: The authors declare no conflicts of interest.

References

1. Jafari, H.; Afshar, S. Improved photodegradation of organic contaminants using nano-TiO_2 and TiO_2-SiO_2 deposited on Portland cement concrete blocks. *Photochem. Photobiol.* **2016**, *92*, 87–101. [CrossRef] [PubMed]

2. Papoulis, D.; Kordouli, E.; Lampropoulou, P.; Rapsomanikis, A.; Kordulis, C.; Panagiotaras, D.; Theophylaktou, K.; Stathatos, E.; Komarneni, S. Characterization and photocatalytic activities of fly ash-TiO_2 nanocomposites for the mineralization of azo dyes in water. *J. Surf. Interfaces Mater.* **2014**, *2*, 261–266. [CrossRef]

3. Chen, X.; Mao, S.S. Titanium dioxide nanomaterials: Synthesis, properties, modifications and applications. *Chem. Rev.* **2007**, *107*, 2891–2959. [CrossRef]

4. Zhang, X.; Pan, J.H.; Du, A.J.; Fu, W.; Sun, D.D.; Leckie, J.O. Combination of one-dimensional TiO_2 nanowire photocatalytic oxidation with microfiltration for water treatment. *Water Res.* **2009**, *43*, 1186. [CrossRef] [PubMed]

5. Canterino, M.; Somma, I.D.; Marotta, R.; Andreozzi, R.; Caprio, V. Energy recovery in wastewater decontamination: Simultaneous photocatalytic oxidation of an organic substrate and electricity generation. *Water Res.* **2009**, *43*, 2710–2716. [CrossRef] [PubMed]

6. Bhatkhande, D.S.; Pangarkar, V.G.; Beenackers, A.A.C.M. Photocatalytic degradation for environmental applications—A review. *J. Chem. Technol. Biotechnol.* **2002**, *77*, 102–116. [CrossRef]

7. Huesken, G.; Hunger, M.; Brouwers, H.J.H. Experimental study of photocatalytic concrete products for air purification. *Build Environ.* **2009**, *44*, 2463–2474. [CrossRef]

8. Lee, B.Y.; Jayapalan, A.R.; Bergin, M.H.; Kurtis, K.E. Photocatalytic cement exposed to nitrogen oxides: Effect of oxidation and binding. *Cem Concr. Res.* **2014**, *60*, 30–36. [CrossRef]

9. Maggos, T.; Plassais, A.; Bartzis, J.G.; Vasilakos, C.; Moussiopoulos, N.; Bonafous, L. Photocatalytic degradation of NO_x in a pilot street canyon configuration using TiO_2-mortar panels. *Environ. Monit. Assess.* **2008**, *136*, 35–44. [CrossRef]

10. Chen, M.; Chu, J. NO_x photocatalytic degradation on active concrete road surface-from experiment to real -scale application. *J. Clean. Prod.* **2011**, *19*, 1266–1272. [CrossRef]

11. Lin, H.J.; Yang, T.S.; Hsi, C.S.; Wang, M.C.; Lee, K.C. Optical and photocatalytic properties of Fe^{3+}-doped TiO_2 thin films prepared by a sol-gel spin coating. *Ceram. Int.* **2014**, *40*, 10633–10640. [CrossRef]

12. Cheng, X.; Gotoh, K.; Nakagawa, Y.; Usami, N. Effect of substrate type on the electrical and structural properties of TiO_2 thin films deposited by reactive DC sputtering. *J. Cryst. Growth* **2018**, *491*, 120–125. [CrossRef]

13. Chen, X.B. Titanium dioxide nanomaterials and their energy applications. *Chin. J. Catal.* **2009**, *30*, 830–851. [CrossRef]

14. Sikora, P.; Cendrowski, K.; Markowska-Szczupak, A.; Horszczaruk, E.; Mijowska, E. The effects of silica/titania nanocomposite on the mechanical and bactericidal properties of cement mortars. *Constr. Build Mater.* **2017**, *150*, 738–746. [CrossRef]

15. Sikora, P.; Horszczaruk, E.; Rucinska, T. The effect of nanosilica and titanium dioxide on the mechanical and self-cleaning properties of water-glass cement mortar. *Procedia Eng.* **2015**, *108*, 146–153. [CrossRef]

16. Mendoza, C.; Valle, A.; Castellote, M.; Bahamonde, A.; Faraldos, M. TiO_2 and TiO_2-SiO_2 coated cement: Comparison of mechanic and photocatalytic properties. *Appl. Catal. B* **2014**, *178*, 155–164. [CrossRef]

17. Umar, I.G.; Abdul, H.A. Heterogeneous photocatalytic degradation of organic contaminants over titanium dioxide: A review of fundamentals, progress and problems. *J. Photochem. Photobiol. C* **2008**, *9*, 1–12.

18. Chen, J.; Kou, S.C.; Poon, C. Photocatalytic cement-based materials: Comparison of nitrogen oxides and toluene removal potentials and evaluation of self-cleaning performance. *Build Environ.* **2011**, *46*, 1827–1833. [CrossRef]

19. Ye, Q.; Mo, R.H.; Yu, Y.C.; Li, G.H.; Huang, Z.Z. Application of polymer cement mortar modified with nitrogen-doped nano-TiO_2 photocatalytic material. *New Build. Mater.* **2009**, *36*, 15–17.

20. Gao, J.W.; Yang, H.; Shen, Q.H. Research progress of TiO_2 in green building materials. *J. Ceram.* **2007**, *28*, 237–239.

21. Wang, C.M.; Shi, H.S.; Li, Y. Research progress of nano-TiO_2 photocatalytic functional building materials. *New Chem. Mater.* **2011**, *39*, 10–12.

22. Dong, R.; Shen, W.G.; Zhong, J.B.; Liao, G.; Chen, H.; Tan, Y. Research progress of photocatalytic self-cleaning concrete. *Concrete* **2011**, *8*, 62–65.

23. Meng, T.; Yu, Y.; Qian, X.; Zhan, S.; Qian, K. Effect of nano-TiO_2 on the mechanical properties of cement mortar. *Constr. Build Mater.* **2012**, *29*, 241–245. [CrossRef]

24. Peng, F.P.; Ni, Y.R.; Zhou, Q.; Kou, J.H.; Lu, C.H.; Xue, Z.Z. New g-C_3N_4 based photocatalytic cement with enhanced visible-light photocatalytic activity by constructing muscovite sheet/SnO_2 structures. *Constr. Build Mater.* **2018**, *179*, 315–325. [CrossRef]

25. Bossa, N.; Chaurand, P.; Levard, C.; Borschneck, D.; Miche, H.; Vicente, J.; Geantet, C.; Aguerre-Chariol, O.; Michel, F.M.; Rose, J.; et al. Environmental exposure to TiO_2 nanomaterials incorporated in building material. *Environ. Pollut.* **2016**, *220 Pt B*, 1160–1170. [CrossRef]

26. Xu, J.; Teng, Y.; Teng, F. Effect of surface defect states on valence band and charge separation and transfer efficiency. *Sci. Rep.* **2016**, *6*, 32457. [CrossRef] [PubMed]

27. Teng, F.; Liu, Z.; Zhang, A.; Li, M. Photocatalytic performances of Ag_3PO_4 polypods for degradation of dye pollutant under natural indoor weak light irradiation. *Environ. Sci. Technol.* **2015**, *49*, 9489–9494. [CrossRef]

28. Teng, F.; Chen, M.; Li, N.; Hua, X.; Wang, K.; Xu, T. Effect of TiO_2 surface structure on the hydrogen production activity of the Pt@CuO/TiO_2 photocatalysts for water splitting. *Chem. Cat. Chem.* **2014**, *6*, 842–847.

29. Wang, X.; Maeda, K.; Thomas, A.; Takanabe, K.; Xin, G.; Carlsson, J.M. A metal-free polymeric photocatalyst for hydrogen production from water under visible light. *Nat. Mater.* **2009**, *8*, 76–80. [CrossRef]

30. Yan, H.J.; Yang, H.X. TiO_2-g-C_3N_4 composite materials for photocatalytic H2 evolution under visible light irradiation. *J. Alloys Compd.* **2011**, *509*, 126–129. [CrossRef]

31. Kang, H.W.; Lim, S.N.; Song, D.; Park, S.B. Organic-inorganic composite of g-C_3N_4-$SrTiO_3$: RH photocatalyst for improved H_2 evolution under visible light irradiation. *Int. J. Hydrogen Energy* **2012**, *37*, 11602–11610. [CrossRef]

32. Fu, Q.; Jiu, J.T.; Cai, K.; Wang, H.; Cao, C.B.; Zhu, H.S. Attempt to deposit carbon nitride films by electrodeposition from an organic liquid. *Phys. Rev. B* **1999**, *59*, 1693–1696. [CrossRef]

33. Zhao, H.; Yu, H.; Quan, X.; Chen, S.; Zhao, H.; Wang, H. Atomic single layer graphitic-C_3N_4: Fabrication and its high photocatalytic performance under visible light irradiation. *Rsc. Adv.* **2014**, *4*, 624–628. [CrossRef]

34. Sun, X.D.; Li, Y.Y.; Zhou, J.; Ma, C.H.; Wang, Y.; Zhua, J.H. Facile synthesis of high photocatalytic active porous g-C_3N_4 with $ZnCl_2$ template. *J. Colloid Interface Sci.* **2015**, *451*, 108–116. [CrossRef] [PubMed]

35. Wang, H.; He, W.; Dong, X.A.; Wang, H.; Dong, F. In situ FT-IR investigation on the reaction mechanism of visible light photocatalytic NO oxidation with defective g-C_3N_4. *Sci. Bull.* **2018**, *63*, 117–125. [CrossRef]

36. Cui, W.; Li, J.; Cen, W.; Sun, Y.; Lee, S.C.; Dong, F. Steering the interlayer energy barrier and charge flflow via bioriented transportation channels in g-C_3N_4: Enhanced photocatalysis and reaction mechanism. *J. Catal.* **2017**, *352*, 351–360. [CrossRef]

37. Cui, W.; Li, J.; Dong, F.; Sun, Y.; Jiang, G.; Cen, W.; Lee, S.C.; Wu, Z. Highly effifficient performance and conversion pathway of photocatalytic NO oxidation on SrO@clusters amorphous carbon nitride. *Environ. Sci. Technol.* **2017**, *51*, 10682–10690. [CrossRef]

38. Shi, L. Preparation and Properties of Carbon Nitride-Based Photocatalytic Materials. Ph.D. Thesis, Harbin Institute of Technology, Harbin, China, 2017.

39. Liao, G.; Chen, S.; Quan, X.; Yu, H.; Zhao, H. Graphene oxide modifified g-C_3N_4 hybrid with enhanced photocatalytic capability under visible light irradiation. *J. Mater. Chem.* **2012**, *22*, 2721–2726. [CrossRef]

40. Hao, Q.; Niu, X.; Nie, C.; Hao, S.; Zou, W.; Ge, J.; Chen, D.; Yao, W. A highly efficient g-C_3N_4/SiO_2 heterojunction: The role of SiO_2 in the enhancement of visible light photocatalytic activity. *Phys. Chem. Chem. Phys.* **2016**, *18*, 31410–31418. [CrossRef]

41. Wang, D.; Yang, P.; Hou, P.K.; Zhang, L.N.; Zhang, X.Z.; Zhou, Z.H.; Xie, N.; Huang, S.F.; Cheng, X. Cement-based composites endowed with novel functions through controlling interface microstructure from Fe_3O_4@SiO_2 nanoparticles. *Cem. Con. Com.* **2017**, *80*, 268–276. [CrossRef]

42. Lin, B.; Xue, C.; Yan, X.; Yang, G.; Yang, G.; Yang, B. Facile fabrication of novel SiO_2/g-C_3N_4 core-shell nanosphere photocatalysts with enhanced visible light activity. *Appl. Surf. Sci.* **2015**, *357*, 346–355. [CrossRef]

43. Lei, J.; Ying, C.; Wang, L.; Liu, Y.D.; Zhang, J.L. Highly condensed g-C_3N_4-modified TiO_2 catalysts with enhanced photo degradation performance toward acid orange 7. *J. Mater. Sci.* **2015**, *50*, 3467–3476. [CrossRef]

44. Cao, S.; Low, J.; Yu, J.; Jaroniec, M. Polymeric photocatalysts based on graphitic carbon nitride. *Adv. Mater.* **2015**, *27*, 2150–2176. [CrossRef] [PubMed]

45. Feng, P.; Chang, H.L.; Liu, X.; Ye, S.X.; Shu, X.; Ran, Q.P. The significance of dispersion of nano-SiO_2 on early age hydration of cement paste. *Mater. Des.* **2020**, *186*, 108320. [CrossRef]

46. Shen, W.Z.; Ren, L.W.; Zhou, H.; Zhang, S.; Fan, W. Facile one-pot synthesis of bimedal mesepormm carbon nitride and its function as a lipase immobilization support. *J. Mater. Chem.* **2011**, *21*, 3890–3894. [CrossRef]

47. Wang, X.X.; Wang, S.S.; Hu, W.D.; Cai, J.; Zhang, L.; Dong, L. Synthesis and photocatalytic activity of SiO_2/g-C_3N_4 composite photocatalyst. *Mater. Lett.* **2014**, *115*, 53–66. [CrossRef]

48. Tolman, R.C. The effect of droplet size on surface tension. *J. Chem. Phys.* **2004**, *17*, 333–337. [CrossRef]

49. Khandaker, M.; Anwar, H. Volcanic ash and pumice as cement additives: Pozzolanic, alkali-silica reaction and autoclave expansion characteristics. *Cem. Concr. Res.* **2005**, *35*, 1141–1144.

50. Wang, D.; Yang, P.; Hou, P.K.; Cheng, X. BiOBr@SiO_2 flower-like nanospheres chemically-bonded on cement-based materials for photocatalysis. *App. Surf. Sci.* **2018**, *30*, 539–548.

Article

Inverse Estimation Method of Material Randomness Using Observation

Dae-Young Kim [1], Pawel Sikora [2,3], Krystyna Araszkiewicz [3] and Sang-Yeop Chung [1,*]

[1] Department of Civil and Environmental Engineering, Sejong University, 209 Neungdong-ro, Gwangjin-gu, Seoul 05006, Korea; kd2young@gmail.com

[2] Building Materials and Construction Chemistry, Technische Universität Berlin, Gustav-Meyer-Allee 25, 13355 Berlin, Germany; pawel.sikora@zut.edu.pl

[3] Faculty of Civil Engineering and Architecture, West Pomeranian University of Technology Szczecin, Al. Piastow 50, 70-311 Szczecin, Poland; krystyna.araszkiewicz@zut.edu.pl

* Correspondence: sychung@sejong.ac.kr; Tel.: +82-2-6935-2471

Received: 16 April 2020; Accepted: 15 June 2020; Published: 16 June 2020

Abstract: This study proposes a method for inversely estimating the spatial distribution characteristic of a material's elastic modulus using the measured value of the observation data and the distance between the measurement points. The structural factors in the structural system possess temporal and spatial randomness. One of the representative structural factors, the material's elastic modulus, possesses temporal and spatial randomness in the stiffness of the plate structure. The structural factors with randomness are typically modeled as having a certain probability distribution (probability density function) and a probability characteristic (mean and standard deviation). However, this method does not consider spatial randomness. Even if considered, the existing method presents limitations because it does not know the randomness of the actual material. To overcome the limitations, we propose a method to numerically define the spatial randomness of the material's elastic modulus and confirm factors such as response variability and response variance.

Keywords: Bayesian updating; spatial randomness; uncertainty; correlation distance; stochastic field

1. Introduction

Research toward the development and incorporation of new building materials in modern engineering structures in order to meet sustainability goals has gathered substantial attention in recent years. Various new cement-based composites have been investigated including lightweight materials (foamed concrete and lightweight aggregates concretes) [1], pervious concretes [2], nano-modified, and self-cleaning materials [3,4]. However, due to the relatively higher production costs of modern building materials than in the case of conventional ones, it is still imperative to find a solution to support the simulating techniques and their accuracy toward decreasing the number of site trials, thus reducing the costs and environmental impact of material.

In general, structural material analysis can be classified into a deterministic or a probabilistic method [5]. The deterministic method uses finite element analysis (FEA) considering the material properties, geometry, and forces. In other words, the various internal and external factors that can exist in a structure are represented by constants. However, these factors are all assumptions, and in the case of actual structures, it would be more reasonable to assume that the factors have different values depending on the position vector in the structural domain.

The probabilistic method assumes that the structure possesses arbitrary material properties, loads, and geometries, and generates a random sample with some statistical characteristics. A deterministic FEA is repeatedly performed on the generated samples to obtain the characteristic behavior of the structure [6,7]. Structural uncertainty, in terms of numerical considerations, can be classified as intrinsic, measurement, and statistical uncertainty, or uncertainty in the mathematical model [8,9]. In each case, the measurement uncertainty is that involved in establishing uncertainty factors through experiments. Experimental and statistical uncertainty is that obtained due to the lack of data due to limited time and space information. Uncertainty in the mathematical model implies uncertainty due to the difference between the actual model of the uncertainty coefficient and the simulated mathematical model. Intrinsic uncertainty is an uncertainty in the structural material, geometric factors expressing the shape of structures, and applied loads [10,11]. These uncertainties are generally considered, both in the actual behavior of the structure and in the reliability analysis. Among them, the stochastic finite element method (SFEM) is mainly focused on the intrinsic uncertainty with the greatest influence. SFEM is a combination of a stochastic method and FEM. The purpose of SFEM is to estimate the uncertain response variation of the structure with respect to the spatial and temporal randomness of the factors in the structural system [12,13]. SFEM is divided into statistical and non-statistical methods in terms of analytical methodology.

There are many analytical methods such as the K-L Expansion [14], Polynomial Chaos Expansion [15], perturbation methods [16], and the weighted integration method that is based on non-statistical methods [17–22]. However, the non-statistical method is mainly based on the first-order expansion or the second-order series expansion for the main variables and is applicable only when the coefficient of variation (COV) of the stochastic fields is low [23]. In fact, when COV ($=\sigma/\mu$) of the stochastic field is large, it is accurate and shows a significant difference from the Monte-Carlo simulation (MCS).

The MCS, which is a representative statistical method, has the advantage of providing a solution to most stochastic problems. However, to obtain a MCS with high efficiency and accuracy, a proper algorithm and considerable time is also required for analysis. This study focuses on estimating the spatial randomness of materials through observation (partial elastic modulus) and overcoming the limitations of the existing statistical methods. For this purpose, a method to numerically define the spatial randomness of the material's elastic modulus was proposed, and the obtained results were demonstrated using factors such as response variability and response variance.

2. Disadvantage of Current Statistical Methods

2.1. Statistical Methods without Considering Spatial Randomness

In the current statistical method, the target random variable, which is the elastic modulus, is assumed to be a normal distribution ($N(\mu, \sigma)$). In this case, a random process satisfying a specific normal distribution is generated as a pseudo time process (here, the random variable satisfies $r_t = r(t)$, $t \in (0, T)$ and an analysis is performed for generated N constant to satisfy the normal distribution ($N(\mu, \sigma)$).

However, as shown in Figure 1, the factor of the actual material's elastic modulus does not have the same value according to the position vector. Therefore, the current method of using a constant field is different from the actual material, and a random sample of the statistical method should satisfy both functions simultaneously.

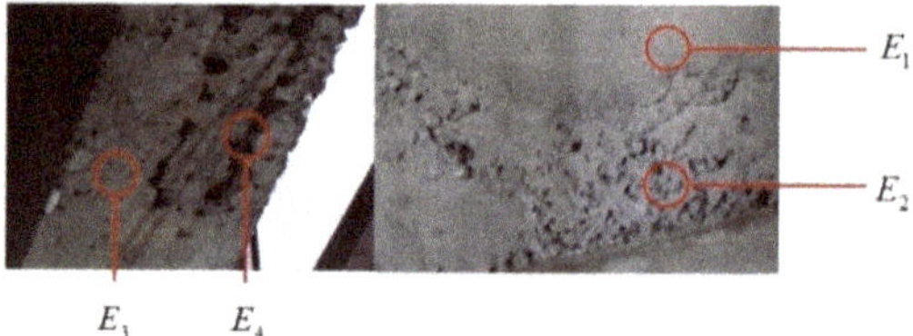

Figure 1. Different elastic modulus in different positions due to material heterogeneity.

2.2. Statistical Methods Considering Spatial Randomness

The spatial randomness distribution of the uncertainty factor is represented by $f(x)$ in the stochastic field function, stochastic field $f(x_i)$, and at any point $x_i \in \Omega_{str}$, belonging to the structural domain Ω_{str}, which must satisfy the probability density function (PDF). At the same time, it must also satisfy the spectral density function (SDF) (or autocorrelation function) representing the aspect existing in the structural domain [24]. Even in the case of a stochastic field with the same mean and standard deviation, the actual shape of the stochastic field can exist in various forms ($\mu_i^k = \mu_j^k$). This variance is due to the characteristics of the stochastic field based on the ensemble concept, which is a statistical characteristic in a direction orthogonal to the stochastic plane.

In general, the distribution characteristic of the stochastic field is determined by the correlation distance. The stochastic field can have two extreme conditions: $f(x; d = 0)$ and $f(x; d = \infty)$. In the former case, the stochastic field appears in the form of a white noise field and contains all the theoretically possible spectra. In the latter, the stochastic field is a constant field with one value, which is called the random variable state [25]. For example, d is infinite when spatial randomness is not taken into consideration.

The distribution of the stochastic field using the correlation distance can simulate (imitate) the spatial randomness of the material. However, it is difficult to determine the spatial randomness of the actual material with this approach. Various methods are available to compute the elastic modulus of concrete, but the spatial randomness is difficult to identify. To investigate the spatial randomness of a target objective, the stochastic field with a correlation distance (d) ranging from zero to infinity needs to be generated. In addition, this spatial characteristic can be examined by the COV trend obtained from the Monte-Carlo simulation, as shown in Figure 2 [26,27].

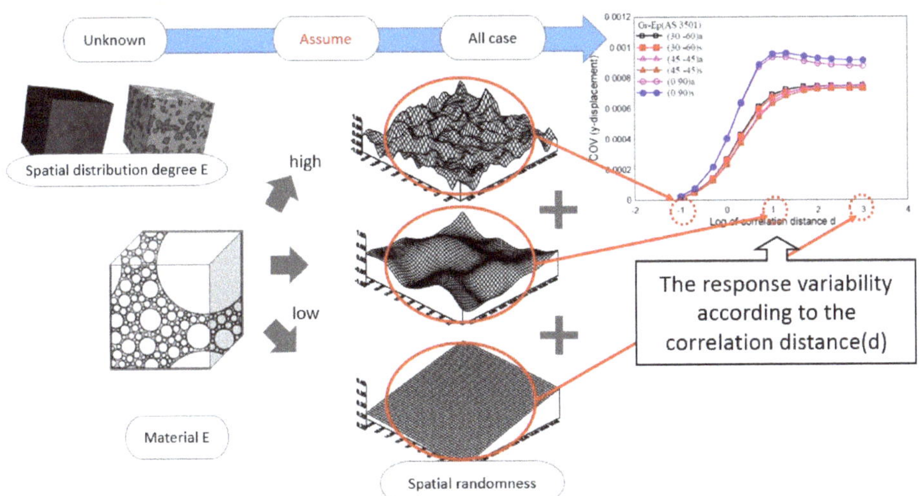

Figure 2. The schematic of the response variability according to the correlation distance (d).

However, since the actual correlation distance (d) of the target elastic modulus cannot be known, the exact result is unknown (which is not a meaningful result). Therefore, we aimed to overcome the problem (disadvantages) of the existing statistical methods.

2.3. Objective and Methods

In this study, we proposed a method of inversely estimating the correlation distance using the observed values and the distance between the observation points. In particular, the population can be predicted using some samples. Then, the obtained real correlation distance d_1 (from the population) is compared with the d_2 (predicted from samples). Once the correlation distance of the actual material is found using the proposed method, the exact behavior of structures by the actual material can be described instead of using the tendency of COV. The process is described in the following three steps. First, generate the stochastic field using the spectral representation method [28,29]. The purpose of creating a sample by applying the algorithm is because the exact d value corresponding to the samples through the algorithm is known. If the proposed inverse estimation method correctly traces d, it starts from the assumption that some data of the actual structure can be used to obtain an actual d corresponding to the structure. Second, we assumed values at some locations to be used as sample data, as shown in Figure 3, and compared the estimated correlation distance (using only some observation) and the correlation distance of the stochastic field. Third, the probability characteristics are obtained using the Bayesian method. This is a method of obtaining new probability characteristics (posterior) by updating the prior probability characteristics based on new test data [30].

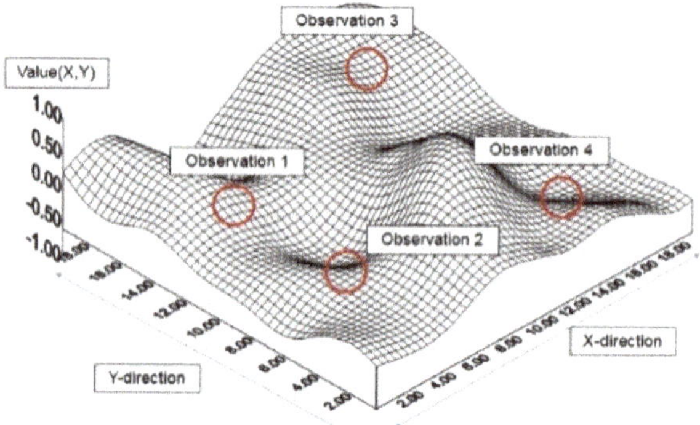

Figure 3. Observation in the stochastic field with spatial randomness.

3. Investigation Procedures and Their Examples

3.1. Consideration Method of Spatial Randomness of Material's Elastic Modulus

For every real-valued 2D-1V homogeneous stochastic field $f_0(x_1, x_2)$ with the mean value equal to zero and a bi-quadrant SDF as $G_{f_0 f_0}(\kappa_1, \kappa_2)$, two mutually orthogonal real processes $u(\kappa_1, \kappa_2)$ and $v(\kappa_1, \kappa_2)$ with orthogonal increments $du(\kappa_1, \kappa_2)$ and $dv(\kappa_1, \kappa_2)$ can be assigned so that:

$$f_0(x_1, x_2) = \int_{-\infty}^{\infty} [\cos(\kappa_1 x_1 + \kappa_2 x_2) du(\kappa_1, \kappa_2) + \sin(\kappa_1 x_1 + \kappa_2 x_2) dv(\kappa_1, \kappa_2)] \tag{1}$$

$$E(x, y) = \overline{E}(1 + f_i(x, y)) \tag{2}$$

$$R(\xi_1, \xi_2) = \sigma_0^2 \exp\left\{-\left(\frac{\Delta \xi_1}{d_1}\right) - \left(\frac{\Delta \xi_2}{d_2}\right)\right\} \tag{3}$$

In the generation of homogeneous random fields, the spectral representation method [31–33] can be employed, which takes advantage of fast Fourier transform. In the random fields of elastic modulus, the stiffness of the elastic modulus is assumed to be homogeneous Gaussian and represented by Equation (2), where $\overline{E}$ is the mean value of the elastic modulus and $f_i(x, y)$ is a homogeneous random field. The auto-correlation functions for the respective random field $f_i(x, y)$ are assumed by Equation (3) [34]. In the spectral representation scheme, the numerical generation of a homogeneous uni-variate i-th random sample $f_i(x, y)$ with a zero mean in two dimensions can be generated via the summation of the cosine functions, as shown in Equation (4) [35,36].

$$f_i(x, y) = \sqrt{2} \sum_{n1=0}^{N_1-1} \sum_{n1=0}^{N_2-1} \left[A_{n1n2} \cos\left(\kappa_{1n_1} x + \kappa_{2n_2} y + \Phi_{n_1 n_2}^{(1)(i)} \right) \right.$$
$$\left. + \widetilde{A}_{n1n2} \cos\left(\kappa_{1n_1} x - \kappa_{2n_2} y + \Phi_{n_1 n_2}^{(2)(i)} \right) \right] \tag{4}$$

$$A_{n1n2} = \sqrt{2 S_{f_0 f_0}\left(\kappa_{1n_1}, \kappa_{2n_2} \right) \Delta \kappa_1 \Delta \kappa_2}, \quad \widetilde{A}_{n1n2} = \sqrt{2 S_{f_0 f_0}\left(\kappa_{1n_1}, -\kappa_{2n_2} \right) \Delta \kappa_1 \Delta \kappa_2} \tag{5}$$

$$\kappa_{1n_1} = n_1 \Delta \kappa_1; \qquad \kappa_{2n_2} = n_2 \Delta \kappa_2 \qquad \Delta \kappa_1 = \kappa_{1u}/N_1; \qquad \Delta \kappa_2 = \kappa_{2u}/N_2 \tag{6}$$

Since the uniform random phase angle $\Phi_{n_1 n_2}$ in Equation (4) is determined depending on n_1 and n_2, it needs to generate two folds of $N_1 \times N_2$ number of values in the range of $[0, 2\pi]$. The upper cut-off limit of wave numbers κ_{1u}, κ_{2u} in the SDF $S_{f_0 f_0}(\kappa_1, \kappa_1)$ can be determined by:

$$\int_0^{\kappa_{1u}} \int_{-\kappa_{2u}}^{\kappa_{2u}} S_{f_0 f_0}(\kappa_1, \kappa_2) d\kappa_1 d\kappa_2 = (1 - \varepsilon) \int_0^{\infty} \int_{-\infty}^{\infty} S_{f_0 f_0}(\kappa_1, \kappa_2) d\kappa_1 d\kappa_2 \tag{7}$$

where ε is set to be 0.001(0.1%) in the numerical generation.

In the basic spectral representation method, as shown in Figure 4, 2D-1V can be expressed as a stochastic field, depending on the mean, standard deviation, and the correlation distance between variables [37].

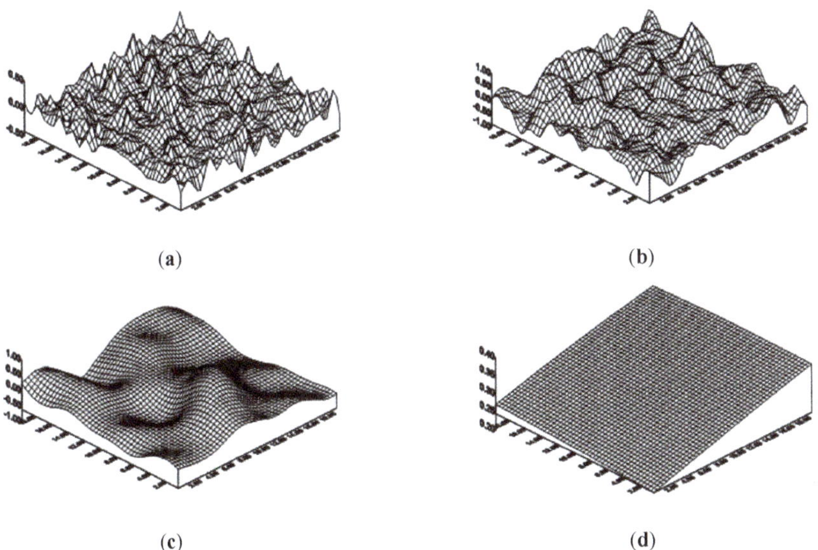

(a)

(b)

(c)

(d)

Figure 4. Stochastic field (2D-1V): (**a**) d = 1.0; (**b**) d = 5.0; (**c**) d = 10.0; (**d**) d = 50.0.

3.2. Bayesian Method

Bayesian inference is a method of statistical inference in which Bayes' theorem is used to update the probability for a hypothesis as more evidence or information becomes available. Bayesian inference is an important technique in statistics, especially in mathematical statistics. Bayesian updating is particularly important in the dynamic analysis of a sequence of data.

Bayesian inference derives the posterior probability as a consequence of two antecedents: a prior probability and a "likelihood function" derived from a statistical model for the observed data. Bayesian inference computes the posterior probability according to Bayes' theorem:

$$posterior \propto likelihood \times prior \tag{8}$$

$$p(\theta|data) \propto p(data|\theta) \times p(\theta) \tag{9}$$

where θ stands for any hypothesis whose probability may be affected by data (called evidence below) [38,39]. Often, there are competing hypotheses, and the task is to determine which is the most probable. $p(\theta)$, the prior probability, is the estimate of the probability of the hypothesis θ before the "*data*," the current evidence, is observed. The evidence "*data*" corresponds to new data that was not used in computing the prior probability.

$p(\theta|data)$, the posterior probability, is the probability of θ given "*data*", after the "*data*" are observed. This is what we want to know: the probability of a hypothesis given the observed evidence. $p(data|\theta)$ is the probability of observing "*data*" given θ, and is called the likelihood. The likelihood function is a function of the evidence, data, while the posterior probability is a function of the hypothesis, θ and can be expressed by the process as shown in Figure 3 [40].

$$\begin{aligned} p(\mu, \sigma^2) &\propto \sigma^{-1}(\sigma^2)^{-(v_0/2+1)} \exp\left[-\tfrac{1}{2\sigma^2}\{v_0\sigma_0^2 + \kappa_0(\mu - \mu_0)^2\}\right] \\ &\times (\sigma^2)^{-n/2} \exp\left[-\tfrac{1}{2\sigma^2}\{(n-1)s^2 + n(\bar{y} - \mu)^2\}\right] \end{aligned} \tag{10}$$

$$\mu_n = \frac{\kappa_0}{\kappa_0 + n}\mu_0 + \frac{n}{\kappa_0 + n}\bar{y} \tag{11}$$

$$\kappa_n = \kappa_0 + n; \, v_n = v_0 + n; \quad \kappa_0 = s^2/\sigma_0^2 \tag{12}$$

$$v_n\sigma_n^2 = v_0\sigma_0^2 + (n-1)s^2 + \frac{\kappa_0 n}{\kappa_0 + n}(\bar{y} - \mu_0)^2 \tag{13}$$

It is expressed as the product of likelihood and prior probability if expanded to a probability distribution, as shown in Equation (10) (posterior probability) [41]. Here, μ_0, μ_n denotes the prior/posterior mean, σ_0, σ_n denotes the prior/posterior standard deviation, and v_0, v_n denotes the prior/posterior degree of freedom. Similarly, $\bar{y}$, s, and n are the test mean, test standard deviation, and test count, respectively, where $s^2 = \left((y_1 - \bar{y})^2 + \cdots + (y_n - \bar{y})^2\right)/n$.

3.3. Consideration of Inverse Estimation (Prediction) Method

For visualization purposes, we considered only 1-axis direction (y = constant or x = constant) of the 2D-1V stochastic field $f_0(x)$, as shown in Figure 5a. The method of estimating the correlation distance, expressing the spatial distribution state, is as follows [42]. First, within the stochastic field, each random variable has an irregular value. Second, the difference in value between two consecutive random variables also has an irregular value, but when d = 1, it will be larger than d = 10. Similarly, as shown in Figure 5c, it is assumed that the non-continuous random variables are also the same. Third, if the two random variables are equal in length, the maximum difference of the value will be similar in the stochastic field, regardless of the position.

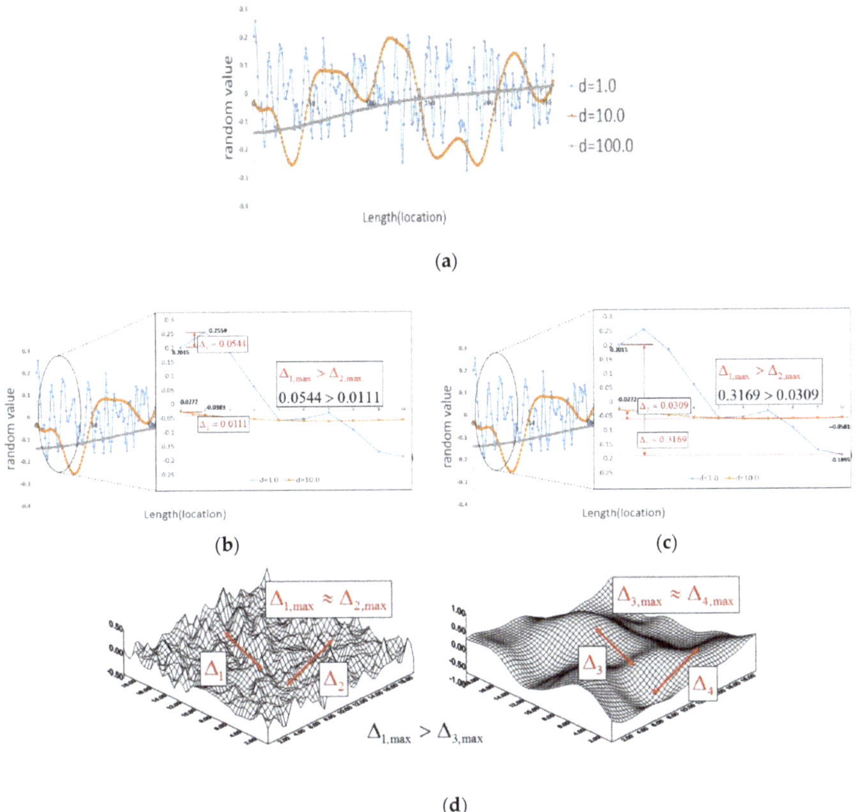

Figure 5. Difference of value between random variables: (**a**) Stochastic fields (1D-1V); (**b**) Difference of value in two continuous variables; (**c**) Difference of value in two non-continuous variables; (**d**) Comparison in stochastic fields with different correlation distance.

4. Verification of the Proposed Method

4.1. Example Model and Method

In this study, the main objective was to find the coefficient (=d) representing the spatial randomness of the elastic modulus using experimental data (=observation) obtained by measuring specific parts of actual samples. However, it is not easy to accurately determine the spatial randomness of material through the measurement of the actual material. This is because there is almost no observed data of the actual structure. Therefore, as mentioned in Section 2, the population was predicted using some samples. Then, we compared the obtained real correlation distance d_1 from the population with the d_2 predicted from samples. We simulated the samples through an algorithm that knows the correlation distance (d) value correctly. The process is described below.

First, as shown in Figure 6, 3000 stochastic fields are generated. Then, 100 out of the 3000 stochastic fields are selected, and the user arbitrarily determines a specific position of the stochastic field, and random variables at that position are collected. At this time, it is assumed that generated stochastic fields (of 3000) are the population measured using a relatively small number of samples (100 samples in this study). In addition, the collected data are assumed as the observation data. Third, the correlation distance is estimated using the collection of data and the Bayesian method; and the statistical characteristics (μ, σ) are obtained. Finally, the statistical characteristics of 3000 stochastic fields are compared with the updated statistical characteristics (and to compare the actual field d_1 with estimated d_2). To verify the performance of the prediction, five cases with different conditions ($d = 1, 5, 10, 20, 50$) were selected and examined.

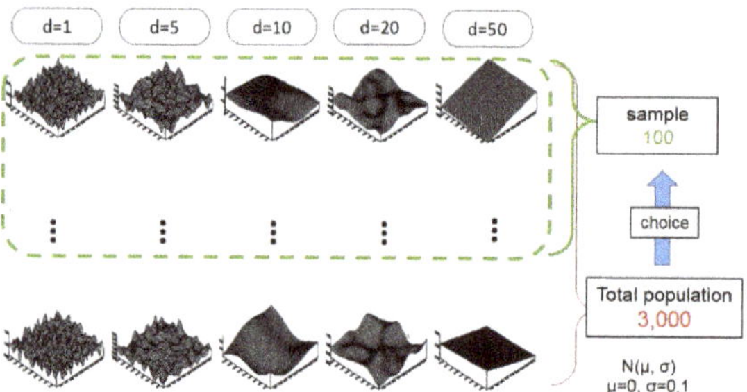

Figure 6. Procedures for estimation (sampling method).

The spatial randomness of the elastic modulus of the target material can be expressed by Equation (2) by applying the spectral representation method to generate stochastic fields $f_i(x, y)$ of 2D-1V. Here, it was assumed that $f_i(x, y)$ had a mean value of 0, a standard deviation of 0.1, and the probability density was assumed to be a normal distribution. $\overline{E}(= 1.0)$ denotes the mean modulus of elasticity and i (1~3000) denotes the number of stochastic fields. Then, we assumed the randomness of the elastic modulus using Equation (2).

The total number of element e of sample fields was 60 elements in the x-direction and 20 elements in the y-direction, as shown in Figure 7a However, we did not know the number of elements of actual material. Therefore, the correlation distance could be found without knowing the number of elements as shown in Figure 7b, Table 1, summarizes the measured values in Figure 7a.

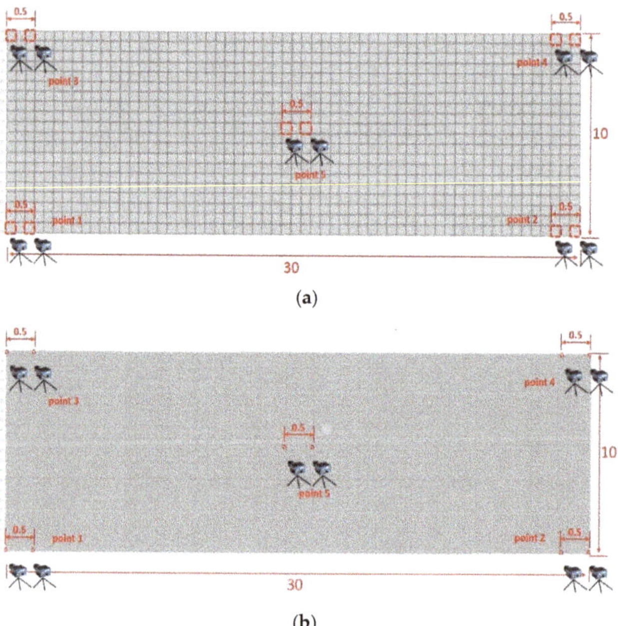

Figure 7. Design domain elements: (**a**) actual material 1200 elements; (**b**) estimated material 6000 elements.

Table 1. Observation data at the distance (= 0.5) per correlation distance.

Max($f_i(x_2)$-$f_i(x_1)$)	d = 1	d = 5	d = 10	d = 20	d = 50
Point 1	0.2211	0.0599	0.0282	0.0119	0.0046
Point 2	0.1745	0.0640	0.0271	0.0117	0.0047
Point 3	0.2241	0.0488	0.0282	0.0118	0.0055
Point 4	0.1631	0.0442	0.0277	0.0116	0.0055
Point 5	0.1832	0.0543	0.0275	0.0117	0.0050

4.2. Results

Part (a) in Figures 8–12 are the stochastic fields composed of material homogeneity (d = 1, 5, 10, 20, 50). These five observation points are assumed to be material samples. The total population was assumed to be 3000, and the design was made so that the true value (elastic modulus) had a mean of 1.0 and a standard deviation of 0.1. The prior probability was assumed to be unknown, with a mean of 0.8 and a standard deviation of 0.2. A total of 100 out of 3000 stochastic fields were chosen continuously from arbitrary positions, and a Bayesian update of five observation points in each stochastic field was performed.

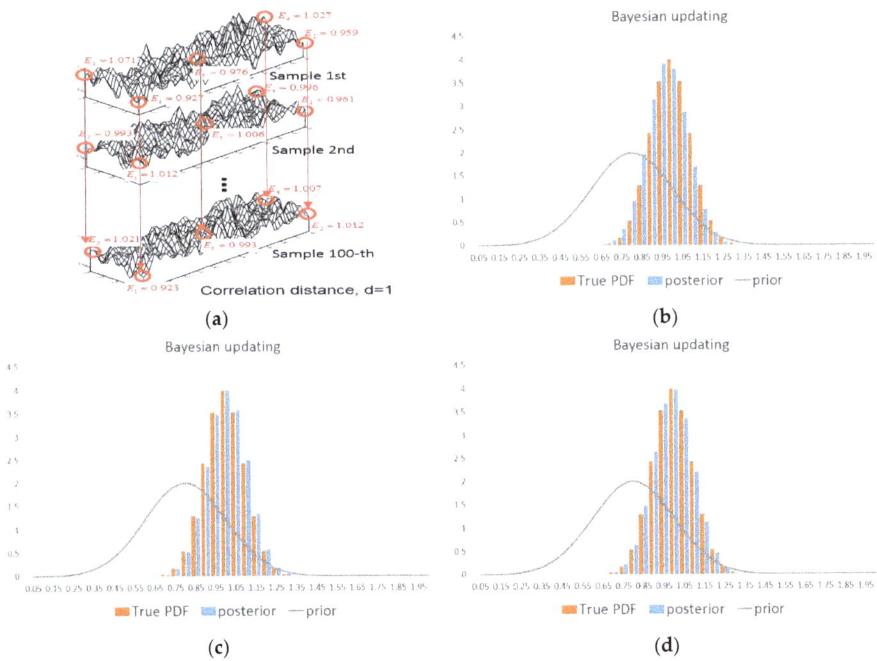

Figure 8. *Cont.*

Figure 8. Bayesian updating at the stochastic field of the correlation distance d = 1.0: (**a**) How to perform Bayesian updating; (**b**) Observation point 1; (**c**) Observation point 2; (**d**) Observation point 3; (**e**) Observation point 4; (**f**) Observation point 5.

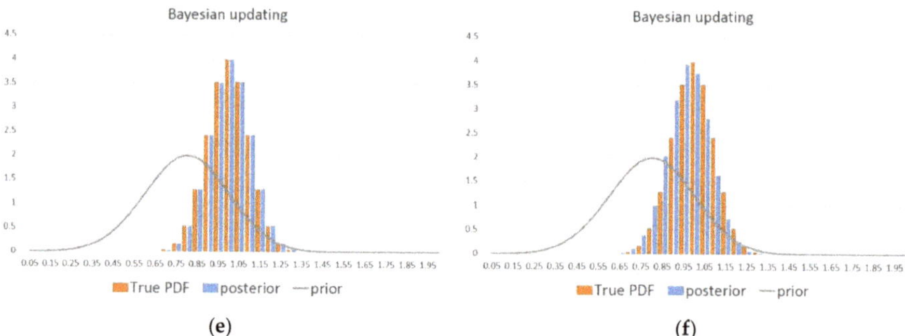

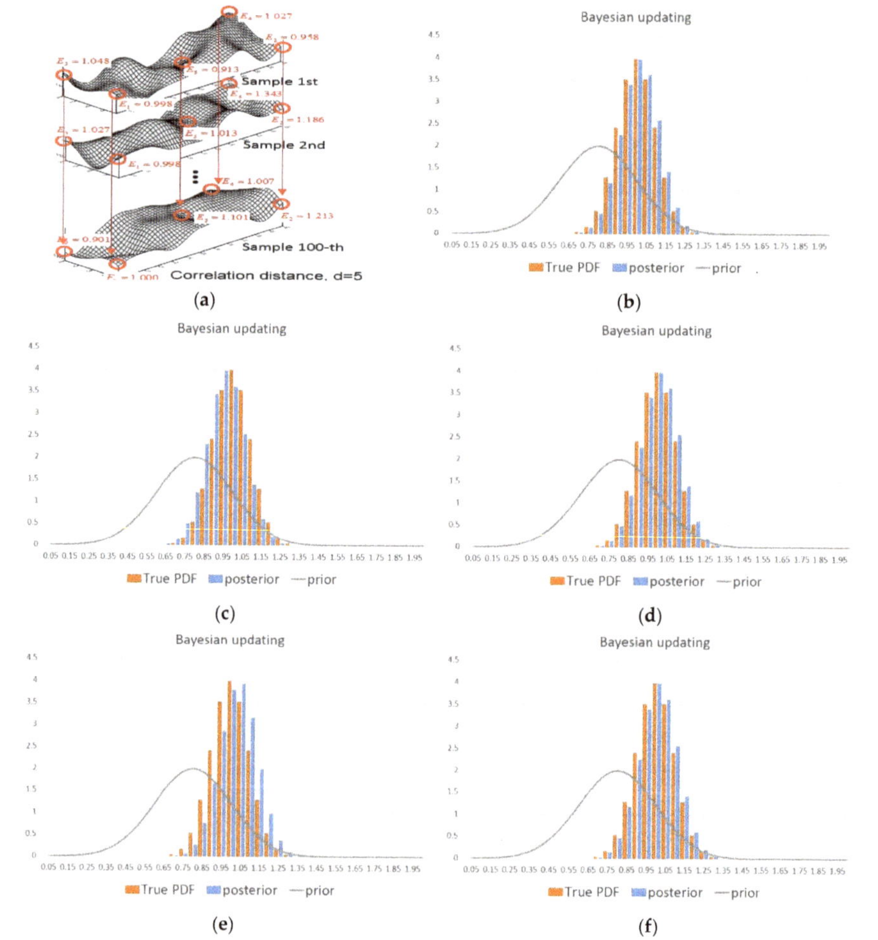

Figure 9. Bayesian updating at the stochastic field of the correlation distance d = 5.0: (**a**) How to perform Bayesian updating; (**b**) Observation point 1; (**c**) Observation point 2; (**d**) Observation point 3; (**e**) Observation point 4; (**f**) Observation point 5.

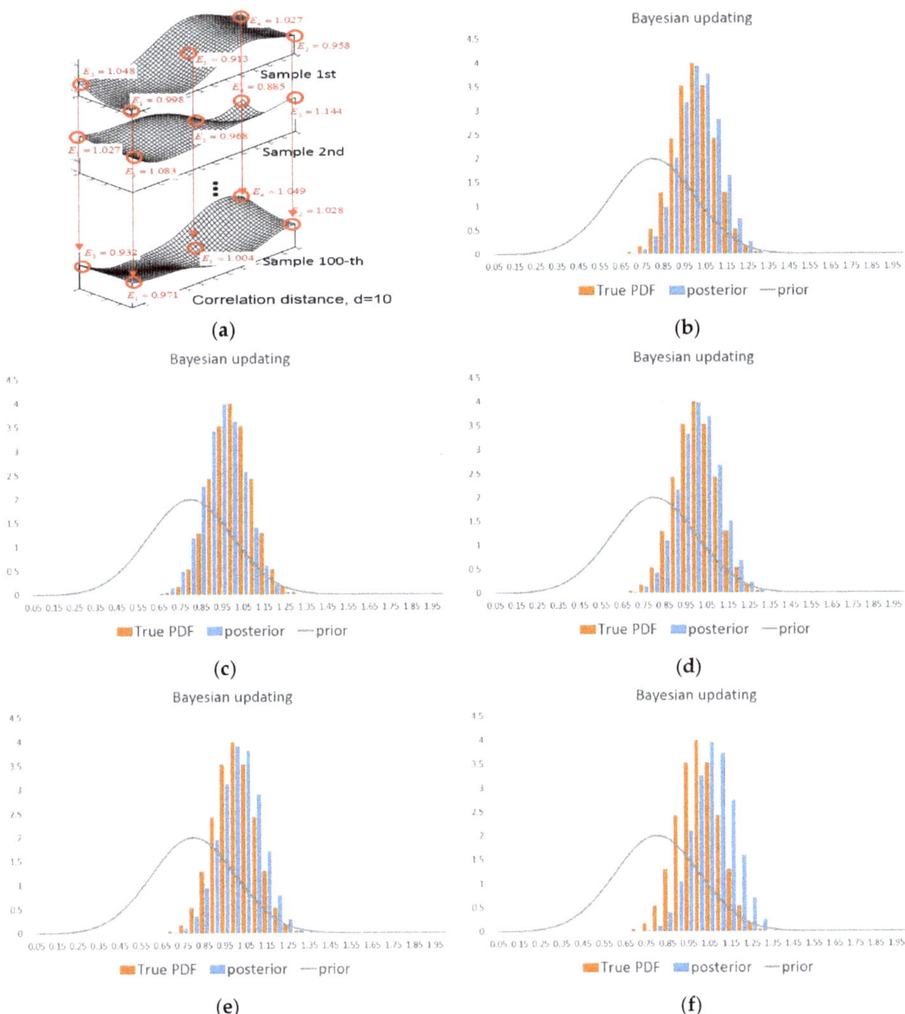

(a)

(b)

(c)

(d)

(e)

(f)

Figure 10. Bayesian updating at the stochastic field of the correlation distance d = 10.0: (**a**) How to perform Bayesian updating; (**b**) Observation point 1; (**c**) Observation point 2; (**d**) Observation point 3; (**e**) Observation point 4; (**f**) Observation point 5.

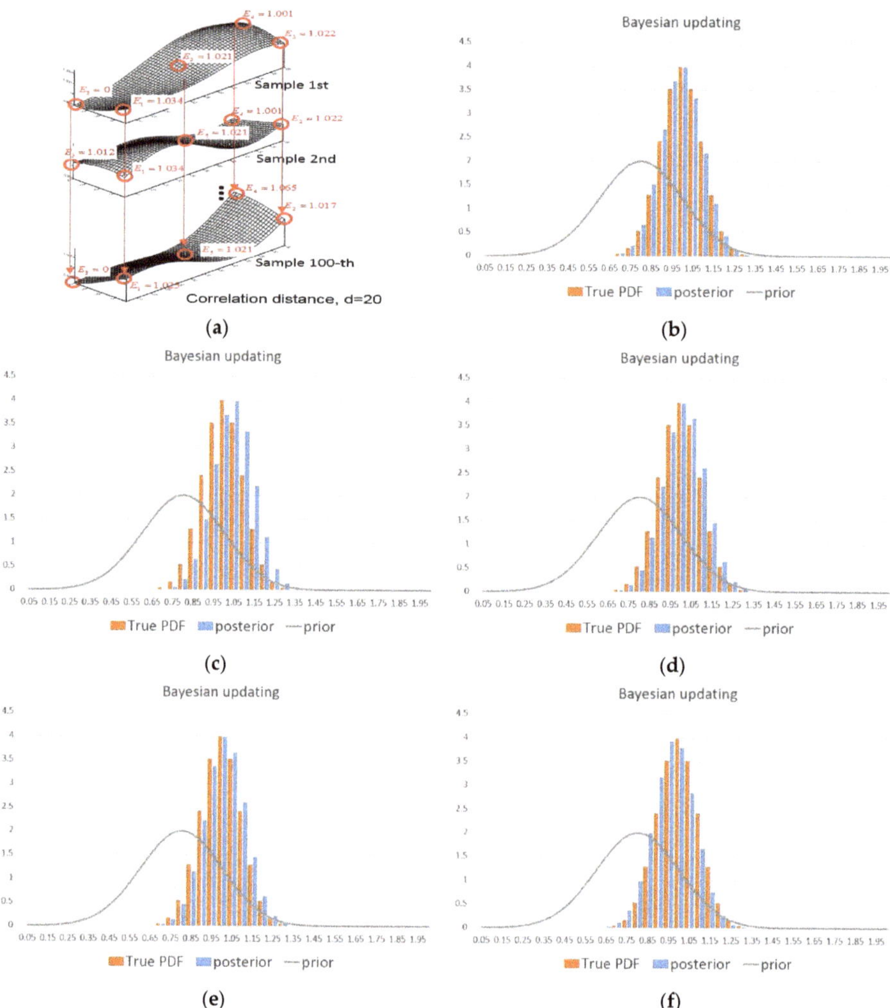

Figure 11. Bayesian updating at the stochastic field of the correlation distance d = 20.0: (**a**) How to perform Bayesian updating; (**b**) Observation point 1; (**c**) Observation point 2; (**d**) Observation point 3; (**e**) Observation point 4; (**f**) Observation point 5.

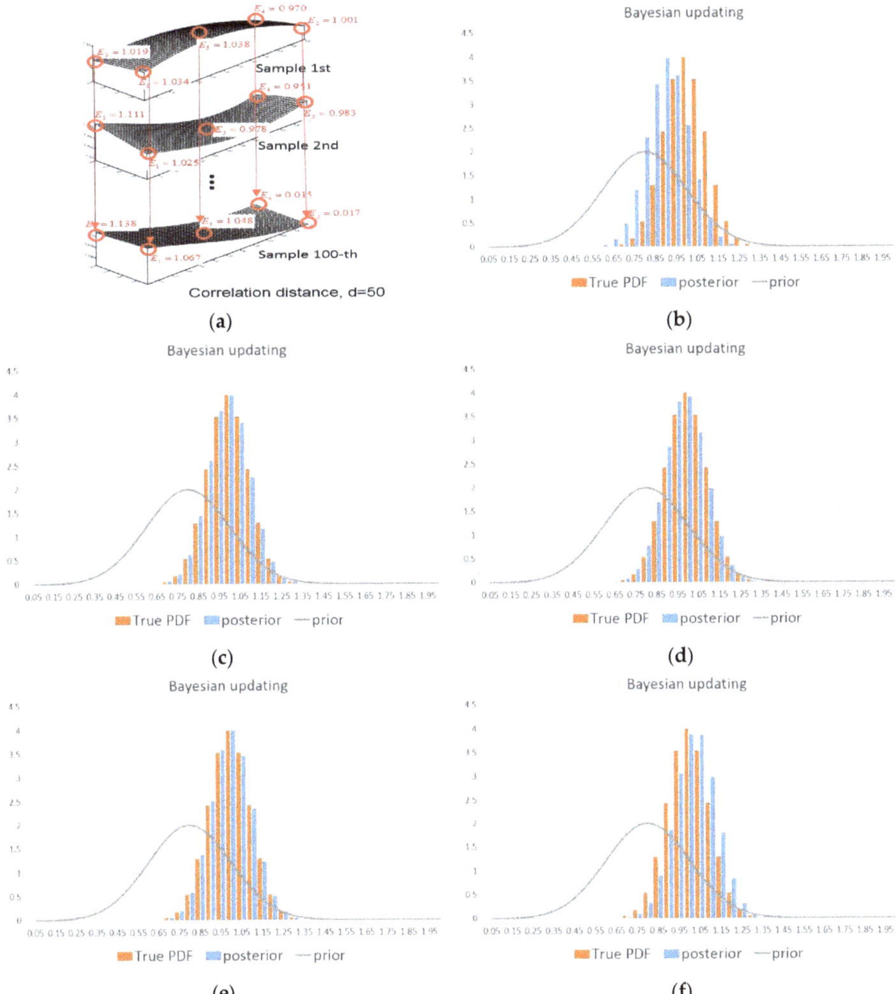

Figure 12. Bayesian updating at the stochastic field of the correlation distance d = 50.0: (**a**) How to perform Bayesian updating; (**b**) Observation point 1; (**c**) Observation point 2; (**d**) Observation point 3; (**e**) Observation point 4; (**f**) Observation point 5.

The elastic modulus values of the observation points (1 to 5) were updated using Bayes' theorem in a direction orthogonal to the stochastic plane. Bayesian updated results for observation points one through five are shown in subimages (b) through (f). The updated posterior probability was similar to the actual value at each position.

Figure 13 shows the comparison between the actual correlation distance and the estimated correlation distance at observation points one to five. Although the errors were different according to the observation positions from (a) to (e), the values achieved between the estimated correlation distance and the actual material showed good agreement. Notably, as the correlation distance (material homogeneity) increased, the correlation distance was accurately detected.

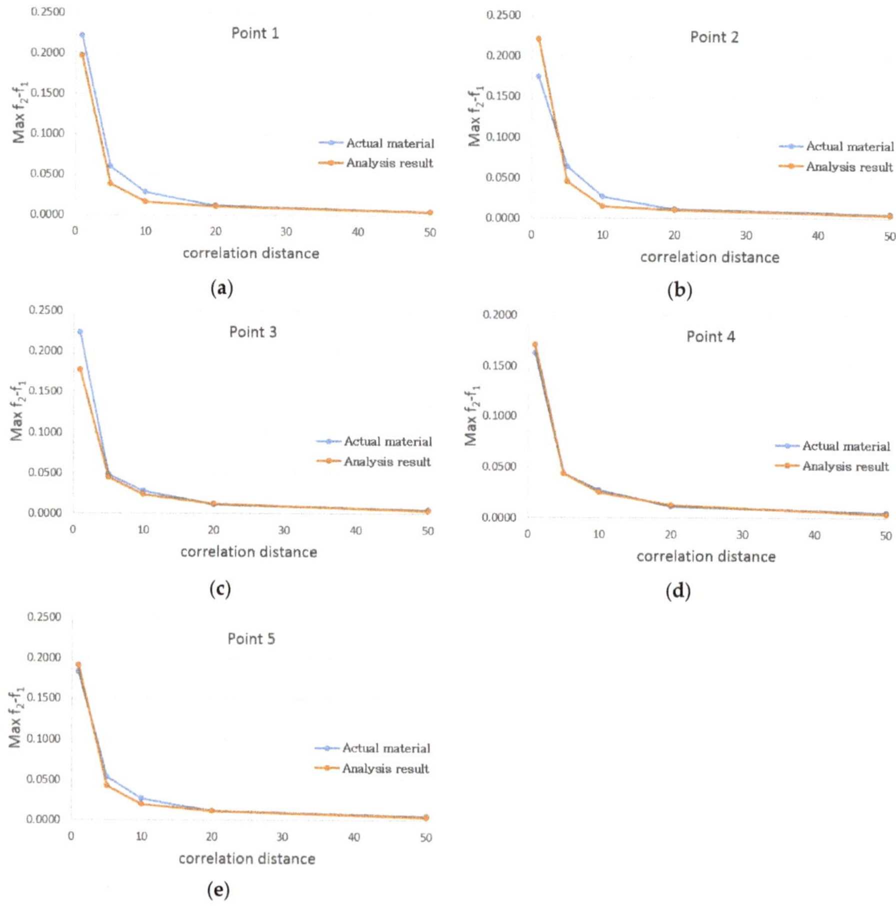

Figure 13. Comparison of d for simulated (imitated) material and inversely analyzed d: (**a**) Observation point 1; (**b**) Observation point 2; (**c**) Observation point 3; (**d**) Observation point 4; (**e**) Observation point 5.

5. Conclusions

The existing design and analysis were based on the deterministic assumption that the design variables had the same value (=constant field) in the structural domain. However, many factors (Poisson's ratio, thickness, etc.) of the actual structure possess temporal and spatial randomness, and the material's elastic modulus contains uncertainties. The results can differ depending on the correlation distance as in the response variance. If the value d in the actual material is not $d = \infty$, the result will be an under or overestimation.

In this study, we proposed a method of inversely estimating the correlation distance coefficient, which indicates the degree of uncertainty of the material, using the measured value of the observation data and the distance between measurement points. Probability characteristics and spatial randomness of the actual material can be estimated using some data only.

The concluding remarks of this paper can be summarized as follows:

- A 2D-1V sample data field was generated using a spectral representation method;
- Correlation distance estimation was performed using only partial values of sample data fields; and
- A new stochastic field was generated that considers the Bayesian updated partial values.

The current probabilistic analysis can simulate spatial randomness through the correlation distance, but it does not know the spatial randomness of the actual material. Therefore, the present results are only assumptions, and their meaning is not conclusive.

In this study, a method of inversely estimating the correlation distance was proposed using measurement (observation), and it was confirmed that the actual correlation distance (d) agreed well with the estimated actual distance. Therefore, we could overcome the current disadvantage and find the true result of the actual structure, and not the assumption. Incidentally, the proposed method has a temporal advantage over the response variance. Verification was performed using an example by assuming that the program running time was 1 min at a time. A total of 100,000 (= 1000 $\times$1 $\times$100) minutes of program running time is required in order to analyze 1000 samples per correlation distance for COV from d = 1~100. This means a decrease in the number of interpretations of the Monte-Carlo simulation, and the program running time was reduced to a total of 1000 (= 1000 $\times$1 $\times$1) minutes.

Future studies need to be conducted considering the application of Poisson's ratio, cross-section (A), thickness (t), and uncertainty factors. Efforts to increase the accuracy of the estimates and actual values are also required in future study.

Author Contributions: Conceptualization, D.-Y.K., P.S., K.A. and S.-Y.C.; Methodology, D.-Y.K. and S.-Y.C.; Software, D.-Y.K. and S.-Y.C.; Validation, D.-Y.K., P.S., K.A. and S.-Y.C.; Formal analysis, D.-Y.K. and S.-Y.C.; Investigation, D.-Y.K. and S.-Y.C.; Resources, S.-Y.C., P.S. and K.A.; Data curation D.-Y.K. and S.-Y.C.; Writing—original draft preparation, D.-Y.K., S.-Y.C., P.S. and K.A.; Writing—review and editing, D.-Y.K., P.S., K.A. and S.-Y.C.; Visualization, D.-Y.K. and S.-Y.C.; Supervision, D.-Y.K. and S.-Y.C.; Project administration, D.-Y.K. and S.-Y.C.; Funding acquisition, D.-Y.K., S.-Y.C. and K.A. All authors have read and agreed to the published version of the manuscript.

Funding: This project was received funding from the European Union's Horizon 2020 research and innovation program under the Marie Sklodowska-Curie Grant agreement no. 841592. Pawel Sikora is supported by the Foundation for Polish Science.

Acknowledgments: The authors would like to acknowledge that this work was supported by the Korea Agency for Infrastructure Technology Advancement (KAIA) and the grant was funded by the Ministry of Land, Infrastructure and Transport (Grant 13IFIP-C113546-01 and Grant 20NANO-B156177-01).

Conflicts of Interest: The authors declare no conflicts of interest.

References

1. Chung, S.-Y.; Abd Elrahman, M.; Kim, J.-S.; Han, T.-S.; Stephan, D.; Sikora, P. Comparison of lightweight aggregate and foamed concrete with the same density level using image-based characterizations. *Constr. Build. Mater.* **2019**, *211*, 988–999. [CrossRef]
2. Pieralisi, R.; Cavalaro, S.H.P.; Aguado, A. Advanced numerical assessment of the permeability of pervious concrete. *Cement Concr. Res.* **2017**, *102*, 149–160. [CrossRef]
3. Janus, M.; Zajac, K. Concretes with photocatalytic activity. In *High Performance Concrete Technology and Applications*; Chapter 7; Yilmaz, S., Ed.; InTech: London, UK; pp. 141–160.
4. Janus, M.; Madraszewski, S.; Zajac, K.; Kusiak-Nejman, E.; Morawski, A.W.; Stephan, D. Photocatalytic activity and mechanical properties of cements modified with TiO$_2$/N. *Materials* **2019**, *12*, 3756. [CrossRef] [PubMed]
5. Schueller, G.I. Developments in Stochastic Structural Methods. *Arch. Appl. Mech.* **2006**, *75*, 755–773. [CrossRef]
6. Nguyen, V.T.; Noh, H.-C. *The Stochastic Finite Element in the Natural Frequency of Functionally Graded Material Beam*; Open Access Books-InthchOpen: London, UK, 2019.
7. Mena, J.D.A.; Margetts, L.; Mummery, P. Practical Application of the Stochastic Finite Element Method. *Arch. Comput. Methods Eng.* **2014**, *23*, 171–190. [CrossRef]
8. Joint Committee on Structural Safety (JCSS). *Probabilistic Assessment of Existing Structures*; RILEM Publications, S.A.R.L., Ed.; The Publishing Company of RILEM: Paris, France, 2001.
9. Wang, W.P.; Chen, G.H.; Yang, D.X.; Kang, Z. Stochastic isogeometric analysis method for plate structures with random uncertainty. *Comput. Aided Geom. Design* **2019**, *74*, 101772. [CrossRef]

10. Choi, C.-K.; Noh, H.-C. Stochastic Finite Element Analysis of Plate Structures by Weighted Integral Method. *Struct. Eng. Mech.* **1996**, *6*, 703–715. [CrossRef]

11. Noh, H.-C. A Formulation for Stochastic Finite Element Analysis of Plate Structures with Uncertain Poisson's Ratio. *Comput. Methods Appl. Mech. Eng.* **2004**, *193*, 4857–4873. [CrossRef]

12. Kiureghian, A.D.; Ke, J.-B. The Stochastic Finite Element Method in Structural Reliability. *Probabilistic Eng. Mech.* **1988**, *3*, 83–91. [CrossRef]

13. Contreras, H. The Stochastic Finite-Element method. *Comput. Struct.* **1980**, *12*, 341–348. [CrossRef]

14. Spanos, P.D.; Beer, M.; Red-Horse, J. Karhunen-Loeve Expansion of Stochastic Processes with a Modified Exponential Covariance Kernel. *J. Eng. Mech. ASCE* **2007**, *133*, 773–779. [CrossRef]

15. Ghanem, R.; Spanos, P.D. Polynomial Chaos in Stochastic Finite Element. *J. Appl. Mech. Trans. ASEM* **1990**, *57*, 197–202. [CrossRef]

16. Kaminski, M. Generalized Perturbation-based Stochastic Finite Element Method in Elastostatics. *Comput. Struct.* **2007**, *85*, 586–594. [CrossRef]

17. Antonio, C.C.; Hoffbauer, L.N. Uncertainty Analysis Based on Sensitivity Applied to Angle-ply Composite Structures. *Reliab. Eng. Syst.* **2007**, *92*, 1353–1362. [CrossRef]

18. Deodatis, G. Weight Integral Method I: Stochastic Stiffness Matrix. *J. Eng. Mech.* **1991**, *117*, 1851–1864. [CrossRef]

19. Lal, A.; Signh, B.N.; Kumar, R. Natural Frequency of Laminated Composite Plate Resting on an Elastic Foundation with Uncertain System Properties. *Struct. Eng. Mech.* **2007**, *27*, 199–222. [CrossRef]

20. Lawrence, M.A. Basis Random Variables in Finite Element Analysis. *Int. J. Numer. Methods Eng.* **1987**, *24*, 1849–1863. [CrossRef]

21. Ngah, M.F.; Young, A. Application of the Spectral Stochastic Finite Element Method for Performance Prediction of Composite Structures. *Compos. Struct.* **2007**, *78*, 447–456. [CrossRef]

22. Papadopoulos, V.; Deodatis, G.; Papadrakakis, M. Flexibility-based Upper Bounds on the Response Variability of Simple Beams. *Comput. Method Appl. Mech. Eng.* **2005**, *194*, 1385–1404. [CrossRef]

23. Micaletti, R.C. Direct Generation of Non-Gaussian weighted Integrals. *J. Eng. Mech. ASCE* **2000**, *126*, 66–75. [CrossRef]

24. Deodatis, G.; Micaletti, R.C. Simulation of Highly Skewed Non-Gaussian Stochastic Processes. *J. Eng. Mech. ASCE* **2001**, *126*, 66–75. [CrossRef]

25. Deodatis, G.; Graham, B.L. A Hierarchy of Upper Bounds on the Response of Stochastic System with Large Variation of Their Properties. *Probabilistic Eng. Mech.* **2003**, *18*, 349–363. [CrossRef]

26. Noh, H.-C. Stochastic Finite Element analysis of Composite Plates Considering Spatial Randomness of Material Properties and Their Correlations. *Steel Compos. Struct.* **2011**, *11*, 115–130. [CrossRef]

27. Noh, H.-C. Stochastic Behavior of Mindlin plate with uncertain geometrical and material parameters. *Steel Probabilistic Eng. Mech.* **2005**, *20*, 296–306. [CrossRef]

28. Shinozuka, M.; Deodatis, G. Simulation of Stochastic Processes by Spectral Representation. *Am. Soc. Mech. Eng.* **1991**, *44*, 191–204. [CrossRef]

29. Shinozuka, M.; Deodatis, G. Simulation of Multi-Dimensional Gaussian Stochastic Fields by Spectral Representation. *Am. Soc. Mech. Eng.* **1996**, *49*, 29–53. [CrossRef]

30. Congdon, P. The Bayesian Method: Its benefits and implementation. In *Bayesian Statistical Modelling*; John Wiley & Sons: Chichester, West Sussex, UK, 2006; pp. 1–18.

31. Carmer, H.; Leadbetter, M. Stationary and Related Stochastic processes. *Soc. Ind. Appl. Math.* **1967**, *10*, 94–95.

32. Svetlisky, V.A. *Stationary Random Function (Processes)*; Springer: Berlin/Heidelberg, Germany, 1962; pp. 77–100.

33. Shinozuka, M.; Deodatis, G. Response Variability of Stochastic Finite Element System. *J. Eng. Mech. ASCE* **1988**, *114*, 499–519. [CrossRef]

34. Shinozuka, M. Monte Carlo Solution of Structural Dynamics. *Comput. Struct.* **1972**, *2*, 855–874. [CrossRef]

35. Ta, D.H.; Nguyen, D.H.; Nguyen, T.K.; Noh, H.-C. The Variability of Dynamic Responses of Beams Resting on Elastic Foundation Subjected to Vehicle With Random System Parameters. *Appl. Math. Model.* **2019**, *67*, 676–687.

36. Ta, D.H.; Lam, N.N. Investigation into the Effect of Random Load on the Variability of Response of Plate by Using Monte Carlo Simulation. *Int. J. Civil Eng. Technol. (IJCIET)* **2016**, *7*, 169–176.

37. Nguyen, H.X.; Ta, D.H.; Lee, J.-H.; Nguyen-Xuan, H. Stochastic buckling behavior of laminated composite structures with uncertain material properties. *Aerosp. Sci. Technol.* **2017**, *66*, 274–283. [CrossRef]

38. An, D.; Won, J.H.; Kim, E.J.; Choi, J.H. Reliability Analysis Under Input Variable and Metamodel Uncertainty Using Simulation Method Based on Bayesian Approach. *Trans. Korean Soc. Mech. Eng. A.* **2009**, *33*, 1163–1170. [CrossRef]
39. Jordan, M.; Kleinberg, M.; Schölkopf, B. Information Science and Statistics: Its benefits and implementation. In *Bayesian Probabilities*; Springer Science: New York, NY, USA, 2006; pp. 24–28.
40. Wei, Z.; Conlon, E.M. Parallel Markov chain Monte Carlo for Bayesian Hierarchical Models with Big Data. in Two Stage. *J. Appl. Stat.* **2019**, *46*, 1917–1936. [CrossRef]
41. Andrew, G.; John, B.C.; Hal, S.S.; David, B.D.; Aki, V.; Donald, B.R. *Bayesian Data Analysis*, 3rd ed.; Taylor & Francis Group: Abingdon, UK, 2013; pp. 29–69.
42. Nguyen, V.T.; Noh, H.-C. Investigation Into the Effect of Random Material Properties on the Variability of Natural Frequency of Functionally Graded Beam. *KSCE J. Civil Eng.* **2017**, *21*, 1264–1272.

Article

Influence of the Aggregate Surface Conditions on the Strength of Quick-Converting Track Concrete

Rahwan Hwang [1], Il-Wha Lee [2], Sukhoon Pyo [3,*] and Dong Joo Kim [1,*]

[1]   Department of Civil and Environmental Engineering, Sejong University, 98 Gunja-Dong, Gwangjin-Gu, Seoul 05006, Korea; yiaamr7304@naver.com

[2]   Korea Railroad Research Institute, 176 Railroad Museum Road, Uiwang-si, Gyeonggi-do 16105, Korea; iwlee@krri.re.kr

[3]   School of Urban and Environmental Engineering, Ulsan National Institute of Sciences and Technology (UNIST), 50 UNIST-gil, Ulju-gun, Ulsan 44919, Korea

*   Correspondence: shpyo@unist.ac.kr (S.P.); djkim75@sejong.ac.kr (D.J.K.)

Received: 25 May 2020; Accepted: 17 June 2020; Published: 24 June 2020

Abstract: This experimental study investigates the effects of the aggregate surface conditions on the compressive strength of quick-converting track concrete (QTC). The compressive strength of QTC and interfacial fracture toughness (IFT) were investigated by changing the amount of fine abrasion dust particles (FADPs) on the aggregate surface from 0.00 to 0.15 wt% and the aggregate water saturation from 0 to 100%. The effects of aggregate water saturation on the compressive strength of the QTC and IFT were notably different, corresponding to the amount of FADPs. As the aggregate water saturation increased from 0 to 100%, in the case of 0.00 wt% FADPs, the IFT decreased from 0.91 to 0.58 MPa·mm$^{1/2}$, and thus, the compressive strength of the QTC decreased from 34.8 to 31.4 MPa because the aggregate water saturation increased the water/cement ratio at the interface and, consequently, the interfacial porosity. However, as the aggregate water saturation increased from 0 to 100%, in the case of 0.15 wt% FADPs, the compressive strength increased from 24.6 to 28.1 MPa, while the IFT increased from 0.41 to 0.88 MPa·mm$^{1/2}$ because the water/cement ratio at the interface was reduced as a result of the absorption by the FADPs on the surface of the aggregates and the cleaning effects of the aggregate surface.

Keywords: Interfacial transition zone; quick-converting track concrete; aggregate surface condition; railway ballast

---

1. Introduction

The traditional track system, a ballasted track, is still widely regarded as one of the favored options for new railway construction projects due to low construction costs and easy maintenance. However, this type of track requires frequent repairs as a result of periodic train loads [1–3]. Lee and Pyo [1] developed a quickly converting track system that converts ballasted railway tracks into concrete tracks using quick-hardening materials.

During the service time of ballasted railway tracks, fine abrasion dust particles (FADPs) of aggregates are generated from the deterioration of aggregates under repeated train loads [2]. Lee et al. [2] experimentally evaluated the influence of FADPs at the interface on the strength of quick-converting track concrete (QTC). They concluded that surface cleaning of aggregates is necessary in order to achieve target strength. Lee et al. [4] additionally assessed the effects of FADPs on interfacial fracture toughness (IFT) between quick-hardening mortar (QM) and ballast aggregates in order to develop a suitable QM with high IFT. They revealed that the use of coarser silica sands and silica fume could produce the required QTC strength with a minimum cleaning process of existing ballast aggregates.

However, the effects of FADPs on the IFT between the aggregate and the QM—and subsequently on the strength of QTC—are not yet fully understood. For instance, ballast aggregates are generally placed in an outside environment and can be easily exposed to water through rain or snow. It is well known that the amount of water in mixing concrete should be adjusted to correspond to the content of water saturation and surface condition of aggregates because the W/C ratio would change if water saturation of the aggregate is not constant [5]. Thus, it can be expected that the water saturation content of the aggregate would substantially affect the compressive strength of QTC.

This experimental study aims to further understand the effects of the aggregate surface condition on the strength development of QTC. The detailed purposes are to experimentally evaluate the effects of FADPs that are adhered to the surface of aggregates and the effects of different water saturation content of the aggregate on the IFT between the aggregate and the QM and, subsequently, on the strength development of QTC.

2. Aggregate Conditions and the Properties of Concrete

Many studies have been carried out to characterize the influence of aggregate conditions, e.g., aggregate moisture content, type, roughness, and surface deformation, on the mechanical properties of concrete [6–18]. Aggregate water saturation generally produces negative effects on the interfacial bonding between the matrix (cement paste or mortar) and inclusions (coarse aggregate) [6,7]. Oliveira and Vazquez [6] investigated the influence of the moisture of recycled aggregate on the strength and durability of concrete and reported that saturated aggregates showed lower flexural strength. Poon et al. [7] revealed that air-dried aggregates produce a better compressive strength than that of water-saturated aggregates. On the other hand, Lee and Lee [8] reported that saturated surface-dry aggregates exhibit higher concrete strength than air-dried and sun-dried aggregates because oil palm shell aggregates are more efficient for internal curing because of the higher absorption capacity of oil palm shell aggregate.

The aggregate type also significantly affects concrete strength [9–13]. Ozturan and Cecen [9] claimed that basalt and limestone generate higher concrete strength than gravel aggregates. The effects of aggregate type on concrete strength show significantly different results in high-strength concrete [10], where crushed quartzite aggregates indicate higher concrete strength than marble aggregates. Beshr et al. [11] also reported that the influences of the type of aggregate on concrete strength are considerable in high-strength concrete. Petros et al. [12] investigated the interpretation of the adverse effects of the secondary products in two types of rocks during their performance as concrete aggregates. They reported that abnormal hydration reactions and considerable swelling of the smectite result in the appearance of defects in the concrete, hence contributing to its low performance. Petros et al. [13] investigated the effects of the aggregate type on concrete strength. They reported that the mineralogy and microstructure of the coarse aggregates affected the strength of concrete.

Aggregate surface roughness [14,15] and the aggregate shapes [16] also have substantial effects on the mechanical properties of concrete. Rao et al. [14] reported that an increase of the roughness of aggregates contributes to an increase in the interfacial bonding between the aggregate and mortar. In addition, Hong et al. [15] revealed that concrete strength variation would correspond to the roughness of the aggregate. Rocco and Elices [16] reported that the concrete that uses crushed aggregates shows better mechanical properties of concrete than the concrete that uses spherical aggregates.

In addition, surface-coating of the aggregate with pozzolanic materials produces noticeable effects on concrete properties [17–19]. Kong et al. [17] concluded that the surface-coating of the aggregate with pozzolanic particles consumes calcium hydroxide (CH) in the pores and interface between the aggregate and mortar, thus forming new hydration products. This phenomenon improves the microstructures around interfacial transition areas, further enhancing the strength and durability of concrete using the recycled aggregate. Choi et al. [19] revealed that the aggregate coated with inorganic powder strengthens the interfacial transition zone, thereby preventing micro-cracking and improving the mechanical performance of the concrete. Petros et al. [20] investigated the effects of three types of recycled materials (beer green glass, waste tile and asphalt) on concrete strength. In addition, the effects of beer green glass with quartz primer and waste tile with quartz primer on the concrete were studied. They reported that the material coated with quartz primer was suitable for obtaining optimal compressive strength results.

Pyo et al. [21] investigated the mechanical properties of ultra high performance concrete (UHPC) incorporating coarser fine aggregates with maximum particle size of 5 mm. They reported that the UHPC mixtures with dolomite and steel fibers with more than one volume percent achieved more than 150 MPa of compressive strength at the age 56 days, and showed strain hardening behavior and limited decrease in tensile strength compared to typical UHPC without coarser fine aggregates.

As many research studies found in the literature point out, the condition of aggregates is a critical factor for the properties of cement-based materials. However, the effects of aggregate conditions on the QTC strength, especially using quick-hardening mortar, are not fully understood. Specifically, investigation is needed of how aggregate water saturation affects the strength development of QTC. It is important to clarify the influence of aggregate water saturation on QTC strength to obtain the target strength of railway tracks using the proposed quick-converting method because ballast aggregates are under various climate conditions.

3. Materials and Methods

Figure 1 illustrates the detailed experimental series for clarifying the effects of aggregate surface conditions on IFT and then on the compressive strength of QTC. As shown in the figure, the first terms, "C" and "F," indicate the compressive and IFT tests, respectively. The second terms ("00" and "15") designate the content (0.00 and 0.15 wt%) of FADPs adhered to the surface of the aggregate with the weight ratio. Furthermore, the last term represents the aggregate water saturation content: "50" indicates a 50% aggregate water saturation content. For example, C-15-100 refers to the compressive test on the specimen using aggregates with 0.15 wt% of the FADP amount and a 100% aggregate water saturation content.

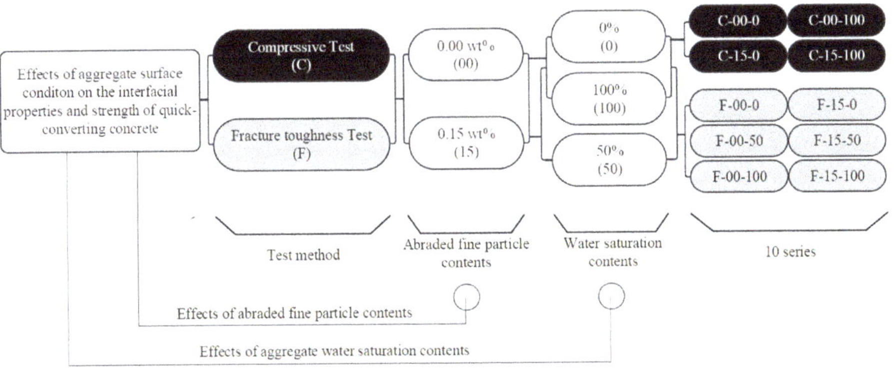

Figure 1. Details of the experimental program.

3.1. Raw Materials and Fabrication

Table 1 shows the composition of the QM matrix. Note that high early strength cement was used in this research for the purpose of the fast construction of the converting track. Polycarboxylic acid superplistizer was used as a high-range water-reducing agent (HRWRA), produced by DongNam Co., Ltd. in South Korea. The setting retarder was provided by SsangYong Co. Ltd. in South Korea. The FADPs were produced containing aggregate powder as by-products during granite processing. The maximum particle size of FADPs is 0.075 mm, which is larger than that of cement (0.005~0.03 mm). The chemical components of the used materials are summarized in Table 2. The fineness of the quick hardening cement is 5400 cm^2/g, which is greater than 3200 cm^2/g for normal cement type I, and the specific gravity is 2.85, which is less than 3.15 for normal cement type I. Both types of sand contained a large amount of SiO_2. The initial setting time of this quick-hardening mortar was 30 min, and more than 30 MPa of compressive strength can be obtained after two hours of curing, as shown in Table 3. It should also be noted that the dosage of the used setting retarder would change, according to the mixing environment and the target curing time.

Table 1. Composition of quick-hardening mortar matrix.

Cement	Water	HRWRA [§]	Sand A [†]	Sand B [‡]	Setting Retarder [*]
1.00	0.40	0.027	0.33	0.34	0.0025

[§] HRWRA: high-range water-reducing agent. [†] Maximum grain size of Sand A: 1.20 mm. [‡] Maximum grain size of Sand B: 0.42 mm. [*] The amount of setting retarder depends on the temperature at mixing.

Table 2. Chemical component of materials.

	Cement	Sand A	Sand B	FADP
SiO_2	13.40	99.0	99.93	72.04
Al_2O_3	15.0	–	0.0313	14.42
Fe_2O_3	1.90	0.12	0.0124	1.22
CaO	51.20	0.35	0.0017	1.82
MgO	1.79	0.35	–	0.71
K_2O	0.43	–	–	4.12
SO_3	12.90	–	–	–
Na_2O	0.13	–	–	3.69
TiO2	–	–	0.0278	0.30
FeO	–	–	–	1.68
P2O5	–	–	–	0.12
MnO	–	–	–	0.05
Ig Loss	3.25	0.20	–	–

Table 3. Compressive strength of mortar.

Type	Specimens	Compressive Strength (MPa)	
		2 h	7 d
	SP1	32.7	58.3
Mortar	SP2	32.3	58.2
	SP3	32.1	57.3
	Average	32.4	57.9

Cylinder specimens (100 mm in diameter and 200 mm in height) were prepared for the compressive QTC strength tests conducted according to ASTM C 39 [22]. QTC specimens were prepared by pre-filling the granite coarse aggregate with a size of 22.4 to 63 mm inside the mold and then injecting the QM. It should be noted that a series of compressive strength tests were carried out because compressive strength is one of the direct indications of overall strength of the ballastless track system. To attach the FADPs onto the aggregate surface, the following steps were conducted: the required weight of FADPs was first measured and mixed with wet coarse aggregate; then, the wet aggregate was fully dried. Saturated aggregates were prepared after immersing the aggregates, in which FADPs were applied according to the parameters in water and removing excess moisture. Although loss of FADPs may occur through this process, it is considered to be a similar condition to the actual track system, e.g., under a meteorological phenomenon.

Figure 2 illustrates the geometry of the IFT samples, and Figure 3 provides photos that show the surface condition of aggregates that correspond to different amounts of FADPs and different aggregate water saturations. The granite aggregate was collected from railway ballast. It can be seen that there is about half of the moisture remaining in the aggregate without FADPs and 50% water saturation of the aggregate surface (Figure 3b). In addition, for the aggregate with 100% water saturation, the entire surface is wet. As the content of aggregate water saturation increases, it can be seen that FADPs are mixed with water on the aggregate surface (Figure 3d–f). All the sliced aggregates were assumed to have identical roughness based on the use of the same cutting method. All specimens were cured at room temperature to create the same environment as the actual track system.

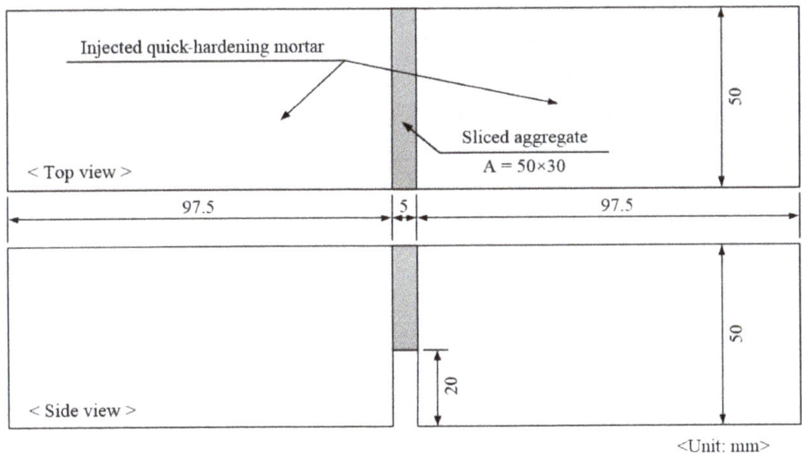

Figure 2. Geometry of the interfacial fracture toughness (IFT) specimens [4].

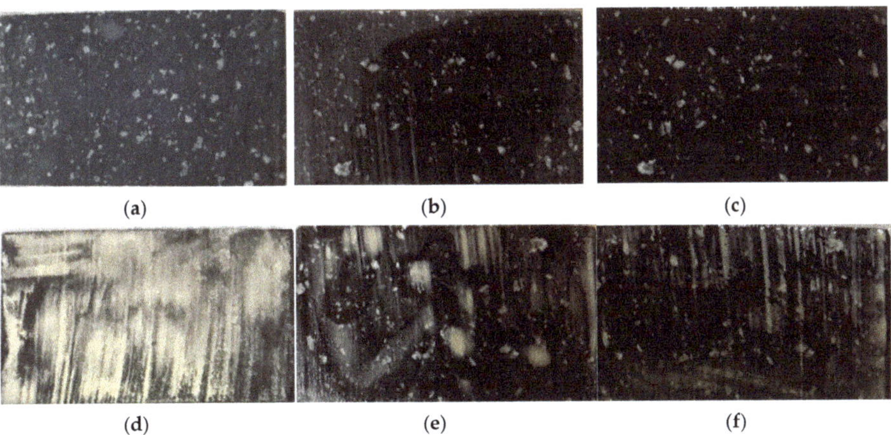

Figure 3. Surface of the aggregates used in this study (**a**) F-00-0, (**b**) F-00-50, (**c**) F-00-100, (**d**) F-15-0, (**e**) F-15-50, (**f**) F-15-100.

The QM mixture was prepared using a laboratory planetary mixer with a 20 L capacity. The mixing process for QM can be found in Lee et al. [4]. The digital image correlation (DIC) method was adopted in this research to characterize the interfacial crack. Random speckles were applied on the surface of IFT samples for the DIC analysis, and Figure 4 shows the prepared IFT samples before and after the stone spray application.

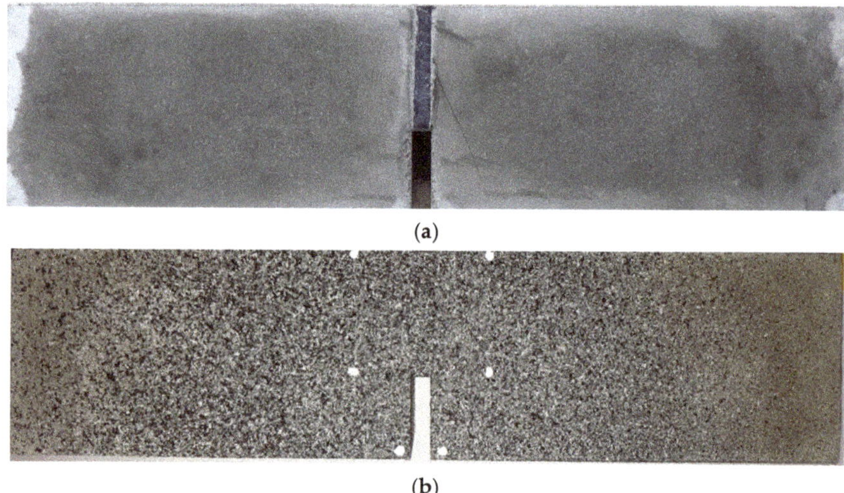

(a)

(b)

Figure 4. Specimen preparation for the digital image correlation (DIC) analysis [4]. (**a**) Surface of the fracture toughness test specimen; (**b**) surface of the fracture toughness test specimen with the applied stone spray.

3.2. Experimental Setup and Procedure

The compressive strength of QTC was measured at the age of 7 d, under a 1.0 mm/min machine displacement rate, using a universal testing machine (UTM) with a 3000 kN capacity. The IFT test was carried out at 7 days of curing, following the same procedure used by Lee et al. [4]. A UTM with a 5 kN capacity was used and the load was applied with a speed of 0.5 mm/min using the displacement control, and the test setup is shown in Figure 5. Images for the DIC analysis were recorded using a high-speed camera and the DIC analysis after testing was performed using commercial DIC software (Tracking Eye Motion Analysis). In both tests, at least three specimens were prepared and tested.

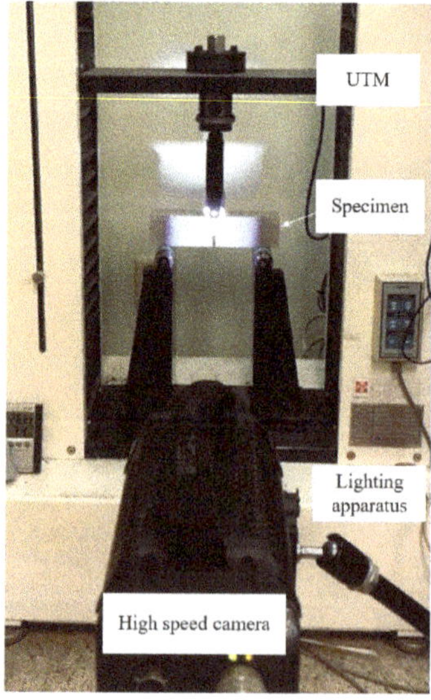

Figure 5. Test setup for the IFT test [4].

4. Results

Table 3 provides the compressive strength of the QM used in this study, while Table 4 provides the strength of QTC that corresponds to different surface conditions of the aggregate. Figure 6 summarizes the influences of contents and of water saturation contents on the IFT. The IFT was calculated using the formula found in Lee et al. [4], which is also given in fracture mechanics by Anderson [23].

Table 4. Compressive strength of quick-converting track concrete.

Test Series	Fine Particle Contents (wt%)	Water Saturation Contents (%)	Compressive Strength (MPa)
C-00-0	0.00	0	34.8
C-00-100		100	31.4
C-15-0	0.15	0	24.6
C-15-100		100	28.1

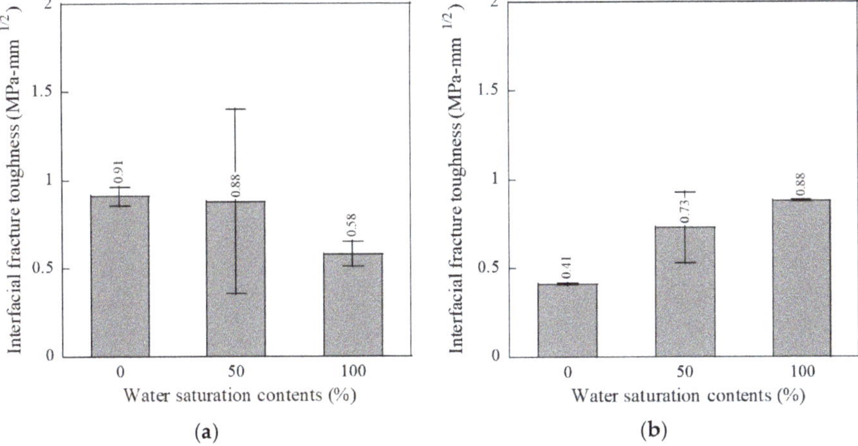

Figure 6. Effects of the water saturation contents on the IFT; (**a**) 0.0 wt% abraded fine particle, (**b**) 0.15 wt% abraded fine particle

4.1. Influence of Fine Abrasion Dust Particles

As the amount of FADPs on the aggregate surface increased from 0.00 to 0.15 wt%, the compressive strength of QTC decreased from 34.8 to 24.6 MPa (Table 4) because the FADPs decreased the IFT between the sliced aggregate and QM. In addition, increasing the amount of FADPs, from 0.00 to 0.15 wt%, resulted in a decrease of IFT, from 0.91 to 0.41 MPa·mm$^{1/2}$, respectively (Figure 6). Although the effects of FADPs on both QTC strength and IFT were similar, the IFT was more sensitive to the FADPs. In Figure 6, the IFT results corresponding to 50% aggregate water saturation showed a large deviation because the location of evaporation of aggregate water saturation would be different. As can be seen in Figure 3b, it is considered that the water saturation on the aggregate surface is inhomogeneous for each specimen, which would result in a high deviation. However, the specimens with 0 and 100% aggregate water saturation produced consistent values of the IFT with smaller deviation because both aggregates with a water saturation of 0% (Figure 3a,d) and aggregate water saturation of 100% (Figure 3c,f) have the same moisture state on the aggregate surface of each specimen. The IFT of the specimens with the aggregate of abraded fine particle contents over 0.20 wt% could not be successfully tested due to premature failure near the interfaces during the casting process and hardening of the specimens. It was also difficult to measure the amount of FADPs less than 0.15 wt%.

Examples of failed surface of the interfaces between the aggregate and QM after the IFT test are shown in Figure 7. F-00-0 without FADPs on the aggregate surface showed no contaminants on the surface of the interface after the test (Figure 7a). On the other hand, as can be seen in Figure 7b, FADPs remained on the surface of the interface after tests, which is a clear indication of the fact that the reduction of IFT was the result of a decrease in IFT between the aggregate and QM. Moreover, Lee et al. [2] concluded that the strength of concrete decreased with increased FADPs because the FADPs deteriorated the interfacial bonding between the aggregate and QM. During the compression tests, the crack propagated in Mode II and/or III as well unlike the Mode I propagation of interfacial cracking during the IFT tests because the shear resistance based on the friction at the interface between the aggregate and matrix influenced the fracture modes. Thus, FADPs at the interface influenced the interfacial friction and consequently deteriorated the compressive strength of the QTC.

(**a**) (**b**)

Figure 7. Failure surface of the interface between the aggregate and the QM after the IFT test (effects of the amount of FADPs); (**a**) F-00-0, (**b**) F-15-0.

4.2. Influence of Aggregate Water Saturation

It is worth noting the different effects of aggregate water saturation on the compressive strength of QTC, which corresponded to the FADP content on the surface of the aggregate. For the specimens without any FADPs, the compressive strength of the QTC decreased from 34.8 to 31.4 MPa (10%) as aggregate water saturation increased from 0 to 100%. However, the compressive strength of the QTC (using the aggregate with 0.15 wt% FADPs) increased from 24.6 to 28.1 MPa (14%) as the aggregate water saturation increased from 0 to 100%.

The influence of water saturation content on the IFT between the aggregate and QM is summarized in Figure 6. The influence of aggregate water saturation on the IFT was similar to the influence on the compressive strength of the QTC. For the samples without any FADPs on the surface of the aggregate, the IFT decreased from 0.91 to 0.58 MPa·mm$^{1/2}$ (36%) as aggregate water saturation increased from 0 to 100%. However, the increase of aggregate water saturation, from 0 to 100%, resulted in an increase of IFT (using the aggregate with 0.15 wt% FADPs) from 0.41 to 0.88 MPa·mm$^{1/2}$ (115%).

Figure 8 shows the fractured interface, after IFT tests, between the aggregate and QM, which corresponds to different water saturations of the aggregate. In the figure, the red boxes indicate the area where pores occurred at the interface. For the specimens without any FADPs on the aggregate surfaces, aggregate water saturation caused an increment of the water/cement ratio of the concrete [5], and consequently, the porosity in the concrete increased [24]. However, for the specimens using the aggregate with 0.15 wt% FADPs, the water/cement ratio at the interface was reduced because of the water absorption by the FADPs on the surface of the aggregates and the cleaning effects of the aggregate surface.

Figure 9 shows photos of the specimens after compressive tests. As shown in Figure 9c, the C-15-0 specimen had a large number of FADPs on the aggregate surface, and thus, porosity occurred at the interface between the coarse aggregate and QM. However, the C-15-100 specimen did not have many FADPs on the aggregate surface, similar to the C-0-100 specimen. All the specimens, except C-15-0, showed similar surface conditions. For the C-15-0 specimen, the interface between aggregate and mortar was clearly separated in comparison with other specimens. Figure 10 illustrates the correlation between the IFT and compressive strength, which indicates that the concrete strength was proportional to the IFT. As the compressive strength of concrete increases, the IFT increases. It is considered that the compressive strength of concrete and IFT are related. However, F(C)-15-100 shows a different tendency. The reason is the moisture and FADPs were removed from interfacial transition zone (ITZ), so that the interfacial bonding was strengthened, but it is considered that the strength of the mortar decreased because the moisture moved to the mortar.

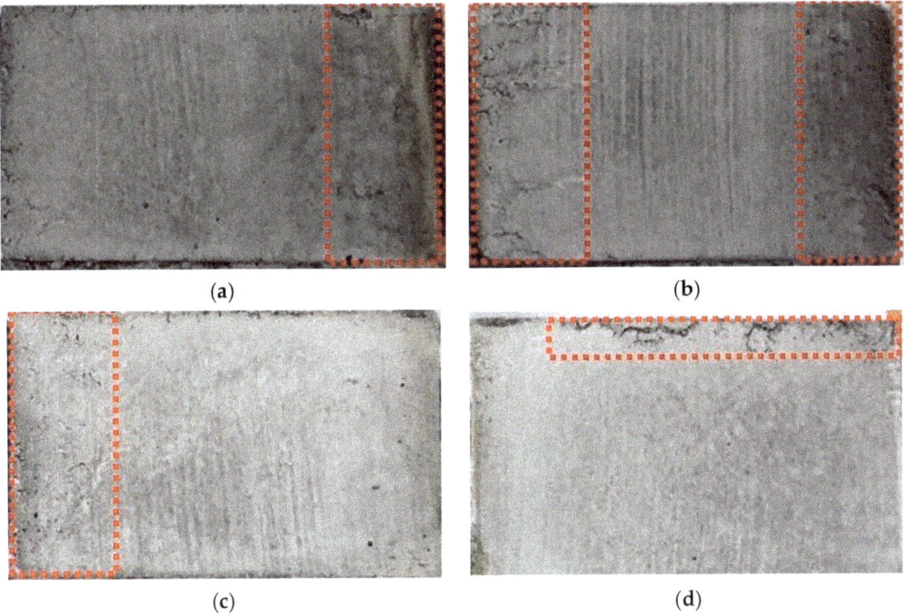

Figure 8. Fractured interface between the aggregate and the QM after the IFT test (effects of the aggregate water saturation content); (**a**) F-0-50, (**b**) F-0-100, (**c**) F-15-50, (**d**) F-15-100.

Figure 11 shows the transverse strain contour during IFT tests. The first set of images shows the state of the specimen under the peak load and the second set of images shows the interfacial crack propagation, taking 0.3 s after peak loads. The last set of images shows the state of the fracture. In the first set of images (peak load), all specimens were unchanged in transverse strain contour. As shown in the second set of images, DIC analysis indicates that the initial crack quickly propagates with decreasing IFT. For the specimens without any FADPs, the crack propagated faster as the contents of aggregate water saturation increased from 0 to 100% (Figure 11a–c). This phenomenon can be explained by the reduced interfacial adhesion due to the increased W/C ratio in ITZ as the aggregate water saturation increased. However, the crack propagation (using the sliced aggregate with 0.15 wt% FADPs) decreased as the content of aggregate water saturation was increased from 0 to 100% (Figure 11d–f). The reduction in the W/C ratio at the interface resulted from the absorption of water by the FADPs on the surface of the aggregate and the cleaning effects of the aggregate surface. It should be noted that crack propagation of the whole series occurred exactly at the interface between the aggregate and QM because the IFT value is significantly low. If it had a high IFT, the adhesion between the aggregate and mortar would have been strengthened, and the crack would have propagated into the mortar. Lee et al. [4] also reported that the higher IFT specimens show that the cracks propagate through the QM.

Figure 9. Failure surface of concrete specimens after compressive tests; (**a**) C-0-0, (**b**) C-0-100, (**c**) C-15-0, (**d**) C-15-100.

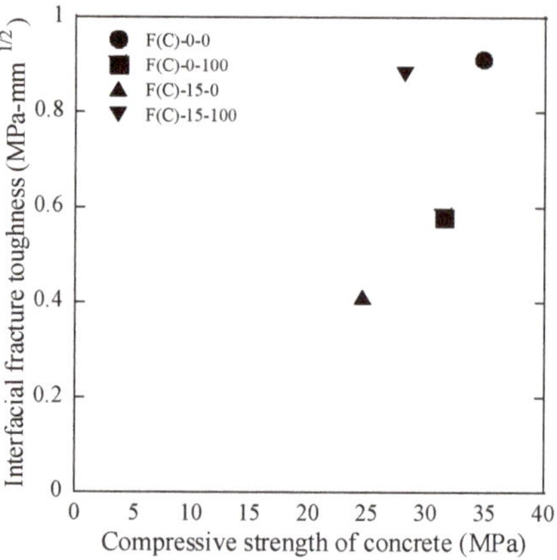

Figure 10. Correlation between the compressive strength and the IFT.

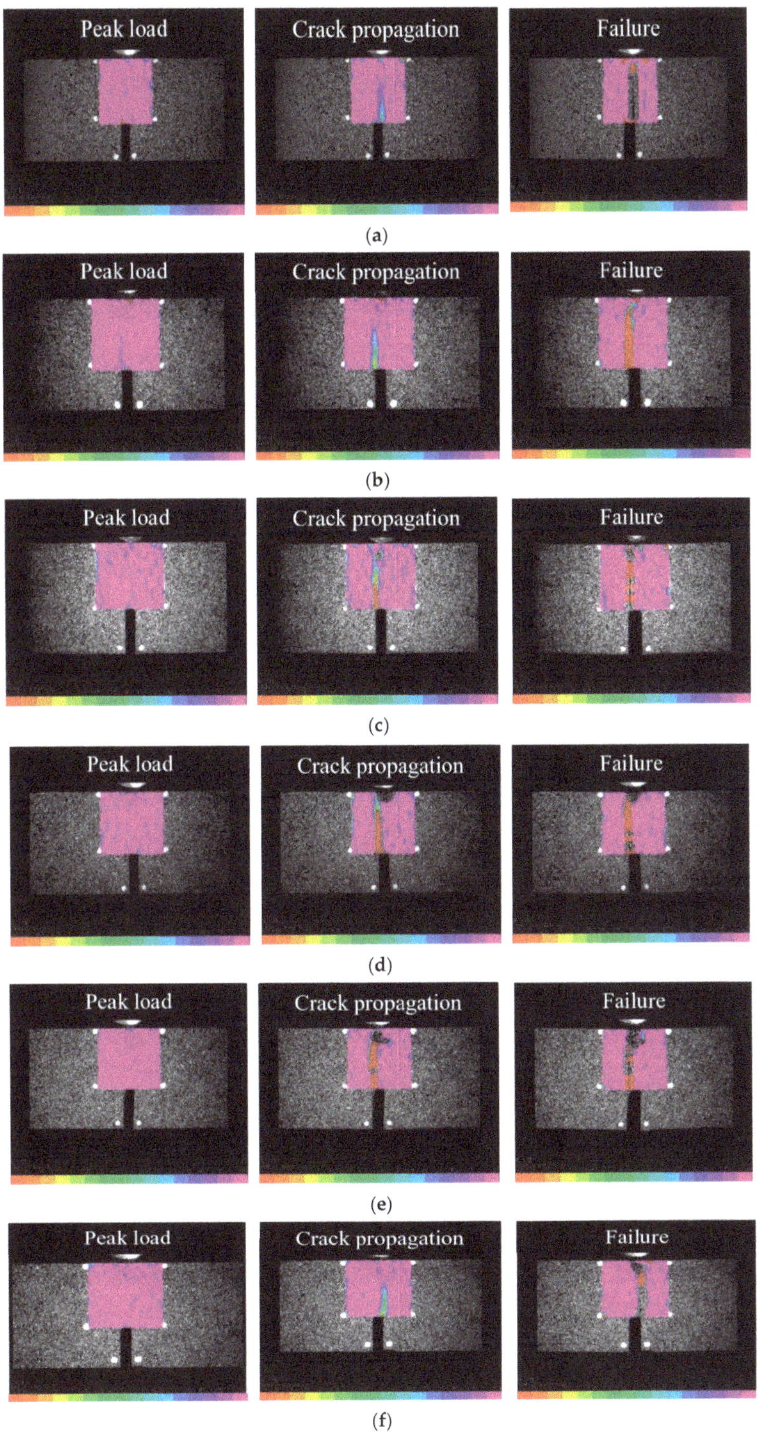

Figure 11. Transverse strain contour of IFT specimens; (**a**) F-00-0, (**b**) F-00-50, (**c**) F-00-100, (**d**) F-15-0, (**e**) F-15-50, (**f**) F-15-100.

5. Conclusions

The compressive strength of QTC and IFT between the aggregates and QM was investigated by changing the amount of FADPs on the surface of the aggregate and the content of aggregate water saturation. The key conclusions can be summarized as:

i. FADPs attached on the aggregate surface deteriorated the interfacial bonding between the aggregate and QM. Accordingly, the compressive strength and IFT decreased as FADPs increased.

ii. For the specimen without any FADPs on the aggregate surface, the IFT decreased as the content of aggregate water saturation increased. The compressive strength of the QTC was also reduced because the porosity at the interface increased due to aggregate water saturation.

iii. However, in the case of 0.15 wt% FADPs on the aggregate surface, both the compressive strength and IFT increased as the content of aggregate water saturation increased. The reduction in the W/C ratio at the interface resulted from the absorption of water by the particles on the surface of the aggregate, and the cleaning effects of the aggregate surface increased the compressive strength as well as IFT.

The results obtained in this study provide fundamental knowledge of the importance of the aggregate surface conditions for the strength development of QTC. Therefore, in order to efficiently apply the quick-converting method, further research is required to improve interfacial bond strength according to aggregate surface conditions. Therefore, research with additional variables is essential. The effects of precipitation on the strength of QTC during the curing period in the actual track system should be investigated in addition to the effects of aggregate water saturation.

Author Contributions: Experiments, Formal analysis, Visualization, Writing original draft, R.H.; I.-W.L.: Supervision, Writing-review and editing, I.-W.L.; Supervision, Formal analysis, Writing-review and editing, S.P.; Supervision, Formal analysis, Writing-review and editing, D.J.K. All authors have read and agreed to the published version of the manuscript.

Funding: The research described herein was sponsored by a grant from the R&D Program of the Korea Railroad Research Institute, Korea. This work was also supported by the National Research Foundation of Korea (NRF) grant funded by the Korea government (MSIT) (No. NRF-2019R1F1A1060906). The opinions expressed in this paper are those of the authors and do not necessarily reflect the views of the sponsors.

Conflicts of Interest: The authors declare no conflicts of interest.

References

1. Lee, I.W.; Pyo, S. Experimental investigation on the application of quick-hardening mortar for converting railway ballasted track to concrete track on operating line. *Constr. Build. Mater.* **2017**, *133*, 154–162. [CrossRef]
2. Lee, I.; Park, J.; Kim, D.J.; Pyo, S. Effects of abraded fine particle content on strength of quick-hardening concrete. *Cem. Concr. Compos.* **2019**, *96*, 225–237. [CrossRef]
3. Lee, I.W.; Pyo, S.; Jung, Y.H. Development of quick-hardening infilling materials for composite railroad tracks to strengthen existing ballasted track. *Compos. Part B Eng.* **2016**, *92*, 37–45. [CrossRef]
4. Lee, I.W.; Hwang, R.; Kim, D.J.; Pyo, S. Interfacial fracture toughness between aggregates and injected quick-hardening mortar. *Cem. Concr. Comp.* **2020**, *106*, 103485. [CrossRef]
5. Kosmatka, S.H.; Kerkhoff, B.; Panarese, W.C. *Design and Control of Concrete Mixtures*, 14th ed.; Pertland Cement Association: Skosie, IL, USA, 2002.
6. de Oliveira, M.B.; Vazquez, E. The influence of retained moisture in aggregates from recycling on the properties of new hardened concrete. *Waste Manag.* **1996**, *16*, 113–117. [CrossRef]
7. Poon, C.S.; Shui, Z.H.; Lam, L.; Fok, H.; Kou, S.C. Influence of moisture states of natural and recycled aggregates on the slump and compressive strength of concrete. *Cem. Concr. Res.* **2004**, *34*, 31–36. [CrossRef]
8. Lee, D.T.C.; Lee, T.S. The effect of aggregate condition during mixing on the mechanical properties of oil palm sheel (OPS) concrete. In *MATEC Web of Conferences*; EDP Sciences: Sarawak, Malaysia, 2017; Volume 87. [CrossRef]

9. Ozturan, T.; Cecen, C. Effects of coarse aggregate type on mechanical properties of concretes with different strengths. *Cem. Concr. Res.* **1997**, *27*, 165–170. [CrossRef]

10. Wu, K.R.; Chen, B.; Yao, W.; Zhang, D. Effect of coarse aggregate type on mechanical properties of high-performance concrete. *Cem. Concr. Res.* **2001**, *31*, 1421–1425. [CrossRef]

11. Beshr, H.; Almusallam, A.A.; Maslehuddin, M. Effect of coarse aggregate quality on the mechanical properties of high strength concrete. *Constr. Build. Mater.* **2003**, *17*, 97–103. [CrossRef]

12. Petrounias, P.; Giannakopoulou, P.P.; Rogkala, A.; Stamatis, P.M.; Tsikouras, B.; Papoulis, D.; Hatzipanagiotou, K. The influence of alteration of aggregates on the quality of the concrete: A case study from serpentinites and andesites from central Macedonia (North Greece). *Geosciences* **2018**, *8*, 115. [CrossRef]

13. Petrounias, P.; Giannakopoulou, P.P.; Rogkala, A.; Stamatis, P.M.; Lampropoulou, P.; Tsikouras, B.; Hatzipanagiotou, K. The Effect of petrographic characteristics and physico-mechanical properties of aggregates on the quality of concrete. *Minerals* **2018**, *8*, 577. [CrossRef]

14. Rao, G.A.; Prasad, B.K.R. Influence of the roughness of aggregate sirface on the interface bond strength. *Cem. Concr. Res.* **2002**, *32*, 253–257. [CrossRef]

15. Hong, L.; Gu, X.; Lin, F. Influence of aggregate surface roughness on mechanical properties of interface and concrete. *Constr. Build. Mater.* **2014**, *65*, 338–349. [CrossRef]

16. Rocco, C.G.; Elices, M. Effect of aggregate shape on the mechanical properties of a simple concrete. *Eng. Fract. Mech.* **2009**, *76*, 286–298. [CrossRef]

17. Kong, D.; Lei, T.; Zheng, J.; Ma, C.; Jiang, J.; Jiang, J. Effect and mechanism of surface-coating pozalanics materials around aggregate on properties and ITZ microstructure of recycled aggregate concrete. *Constr. Build. Mater.* **2010**, *24*, 701–708. [CrossRef]

18. Li, J.; Xiao, H.; Zhou, Y. Influence of coating recycled aggregate surface with pozzolanic powder on properties of recycled aggregate concrete. *Constr. Build. Mater.* **2009**, *23*, 1287–1291. [CrossRef]

19. Choi, H.; Choi, H.; Lim, M.; Inoue, M.; Kitagaki, R.; Noguchi, T. Evaluation on the mechanical performance of low-quality recycled aggregate through interface enhancement between cement matrix and coarse aggregate by surface modification technology. *Int. J. Concr. Struct. Mater.* **2016**, *10*, 87–97. [CrossRef]

20. Petrounias, P.; Giannakopoulou, P.P.; Rogkala, A.; Lampropoulou, P.; Tsikouras, B.; Rigopoulos, I.; Hatzipanagiotou, K. Petrographic and Mechanical Characteristics of Concrete Produced by Different Type of Recycled Materials. *Geosciences* **2019**, *9*, 264. [CrossRef]

21. Pyo, S.; Kim, H.K.; Lee, B.Y. Effects of coarser fine aggregate on tensile properties of ultra high performance concrete. *Cem. Concr. Compos.* **2017**, *84*, 28–35. [CrossRef]

22. ASTM C 39. *Standard Test Method for Compressive Strength of Cylindrical Concrete Specimens*; ASTM International: West Conshohoken, PA, USA, 2018.

23. Anderson, T.L. *Fracture Mechanics Fundamentals and Applications*; Taylor & Francis: Boca Raton, FL, USA, 1991.

24. Sakai, E.; Sugita, J. Composite mechanism of polymer modified cement. *Cem. Concr. Res.* **1995**, *25*, 127–135. [CrossRef]

Article

Determination of Mechanical Characteristics for Fiber-Reinforced Concrete with Straight and Hooked Fibers

Zuzana Marcalikova [1,*], Radim Cajka [1], Vlastimil Bilek [2], David Bujdos [2] and Oldrich Sucharda [2]

[1] Department of Structures, Faculty of Civil Engineering, VSB-Technical University of Ostrava, Ludvíka Poděště 1875/17, 708 00 Ostrava-Poruba, Czech Republic; radim.cajka@vsb.cz

[2] Department of Building Materials and Diagnostics of Structures, Faculty of Civil Engineering, VSB-Technical University of Ostrava, Ludvíka Poděště 1875/17, 708 00 Ostrava-Poruba, Czech Republic; vlastimil.bilek@vsb.cz (V.B.); david.bujdos@vsb.cz (D.B.); oldrich.sucharda@vsb.cz (O.S.)

* Correspondence: zuzana.marcalikova@vsb.cz; Tel.: +42-059-732-1396

Received: 28 April 2020; Accepted: 19 June 2020; Published: 25 June 2020

Abstract: Fiber-reinforced concrete has a wide application in practice, and many fields of research are devoted to it. In most cases, this is a specific problem, i.e., the determination of the mechanical properties or the test method. However, wider knowledge of the effect of fiber in concrete is unavailable or insufficient for selected test series that cannot be compared. This article deals with the processing of a comprehensive test study and the impact of two types of fibers on the quantitative and qualitative parameters of concrete. Testing was performed for fiber dosages of 0, 40, 75, and 110 kg/m^3. The fibers were hooked and straight. The influence of the fibers on the mechanical properties in fiber-reinforced concrete was analyzed by functional dependence. The selected mechanical properties were compressive strength, splitting tensile strength, bending tensile strength, and fracture energy. The results also include the resulting load–displacement diagrams and summary recommendations for the structural use and design of fiber-reinforced concrete structures. The shear resistance of reinforced concrete beams with hooked fibers was also verified by tests.

Keywords: concrete; fiber-reinforced concrete; mechanical characteristics; three-point bending test; fracture energy

---

1. Introduction

One of the main disadvantages of concrete for structure design is its low tensile strength. For this reason, research has focused on improving and developing a new composition and reinforcement that can eliminate this major disadvantage [1,2]. Scattered reinforcement has local tensile effects and affects the spatial stress in the concrete structure. As a result of the use of scattered reinforcement, both the tensile strength and the load–displacement diagram itself are significantly affected. There are many variants of fiber-reinforced concrete, which differ mainly in the material and shape properties of the fibers used [3]. The shape of the fibers affects the mechanical properties [4]. The resulting mechanical properties of fiber-reinforced concrete are further influenced by the recipe of the plain concrete, fiber volume fraction, and processing of the fiber-reinforced concrete (e.g., degree of compaction). The effects include the development of micro-cracks, resistance to temperature changes, shrinkage, and impact strength. The wider development and application of fiber-reinforced concrete have been prevented primarily by the difficultly of correct technological and static assessment. The price of fibers also increases the overall cost of concrete production. On the other hand, the complexity of the work due to the laying of steel bar-reinforced concrete is eliminated. In selected cases of structural design, fiber-reinforced concrete can also replace classical reinforcement. In terms of the technological process, it is especially necessary to ensure that the resulting composite has a homogeneous structure [5]. This is especially true in the case of a higher volume of the fiber fraction, for which processing is difficult [6].

A very important property that is sometimes largely neglected is the density of concrete/fiber-reinforced concrete. We distinguish several types of densities. These are the density of the fresh concrete mixture, the density of the sample in natural storage, the density of the dried sample, and the density of the sample stored in water. The difference between the individual densities can be in the order of a few percent. The density is primarily determined by the concrete recipe, i.e., the type of aggregate used, cement content, and water content (respectively water–cement ratio). In the case of fiber-reinforced concrete, the density is further affected by the volume of the fiber fraction. Another factor that can strongly affect the density is the degree of compaction and the number of pores. In the case of fiber-reinforced concrete, the degree of compaction is greatly influenced by the fibers themselves. It is true that the larger the number of fibers in the concrete, the more difficult it is to perform thorough compaction. The density can also further affect the resulting mechanical properties (compressive strength, tensile strength, and others), the permeability of the concrete, or the resistance to chlorine.

Structural parts suitable for fiber-reinforced concrete include slabs and floors [7–9] or beams without shear reinforcement [10]. The main limitation of fiber-reinforced concrete, however, is that even with very good mechanical properties of compression and tensility, the full replacement of classical reinforcement cannot be expected. However, the ductility and the safety level of a typical reinforced concrete beam and slab remain significantly higher. The optimal solution is to use fiber-reinforced concrete in combination with classical reinforcement. These are mainly lightweight sections without shear reinforcement in the variant with classical or prestressed reinforcement. The design and use of fiber-reinforced concrete must be approached qualitatively because it is a different material to conventional concrete. The fibers affect the mechanical properties [11,12] and fracture properties [13,14], i.e., the deformability and ductility. The biggest difference is in the mentioned tensile strength. However, it is necessary to use a load–displacement diagram to describe tensile strength [15]. This is because it is necessary to differentiate between the phases of behavior before crack formation and during crack formation and fiber activation. We also distinguish the nature of the behavior of the concrete matrix and fibers in crack propagation, namely, whether the matrix cracks, the fibers break, or the fibers are pulled out of the matrix. For practical reasons, these phenomena are described with the help of the residual tensile strength [16] or elastic properties of the material [17]. For concrete of lower compressive and tensile strengths, steel fibers are typically rapidly activated. Furthermore, the crack width increases, and the fibers are pulled out (not broken). The cohesion of concrete with fibers is lower.

On the other hand, in the case of high-strength concrete, the effect of the fibers on the tensile strength is typically less, the fibers are fully activated, and the fibers break when the crack increases. The specific problem in using fiber-reinforced concrete is the recipe design [6,18]. Typically, the volume of the steel fiber fraction is no more than 3–4%. However, the workability of concrete beyond a fiber volume fraction of 2% is very difficult. The presented research deals with fiber volume fractions of 0, 0.51, 0.96, and 1.40%, which correspond to fiber dosages of 0, 40, 75, and 110 kg/m^3, respectively. The fibers also affect the microstructure of the concrete and change the nature of the bonding between the aggregate and the cement. As a result, it is advisable to adjust the composition of the concrete recipe by increasing the fiber volume fraction. A fiber volume fraction above 1%, given the same recipe, can, in some cases, also negatively reduce its compressive strength. Current directions for fiber-reinforced concrete include the use of hybrid fiber-reinforced concrete [19–21], fiber-reinforced self-compacting concrete [22,23], or high-strength concrete [24].

Research into fiber-reinforced concrete and its mechanical properties has led to several recommendations and standards [25–27]. The most important ones are the Model Code 2010 [28], the RILEM recommendations [29], and the German DAfStb standard [30]. In addition to the basic mechanical properties of fiber-reinforced concrete required for the design of structures, it is also important to address durability, which also includes research on chloride in concrete [22].

2. Load–Displacement Diagram and Fracture Energy

For concrete and concrete structures, it is also very important to evaluate fracture-mechanical parameters [31], such as absorbed fracture energy and fracture energy. The fracture parameters need to be determined for both plain concrete and concrete with dispersed reinforcement, where it is necessary to determine the brittleness of this material. These fracture-mechanical parameters also serve very well to predict crack formation [32]. The size of the fracture energy tells us how much energy contributes to the formation of cracks. Fracture tests in a three-point bend of a notched beam are usually used to determine these parameters. The determination of fracture-mechanical properties is also important for the assessment of concrete structures and numerical modeling [33].

Figure 1 describes the difference in the behavior of plain concrete and fiber-reinforced concrete. In the case of plain concrete, the load–displacement diagram after peak strength has been reached can be characterized as an exponential curve. When the tensile strength of plain concrete is reached, the element suddenly breaks. Therefore, this tensile strength is not applied in practice. Adding fibers to the concrete changes the behavior of the element and substantially changes the shape of the load–displacement diagram [34,35]. After reaching the concrete strength (peak strength), there is a macro-crack, which quickly opens and lengthens. At this point, the fibers in the concrete are activated, and the tensile stress transfers these fibers. Tensile behavior can be defined in two ways, i.e., tensile softening [36] and tensile hardening. Whether tensile softening or tensile hardening occurs is influenced mainly by the number of fibers but certainly also by the recipe of the concrete and the processing of the fiber-reinforced concrete. In that case, another peak appears in the load–displacement diagram, which we call the residual tensile strength. When this limit is reached, the tensile capacity is exhausted. The tensile strength of the matrix or the tensile strength of the matrix with fibers is exceeded, and a tensile softening is applied, which is typical for plain concrete or fiber-reinforced concrete with a low fiber volume fraction.

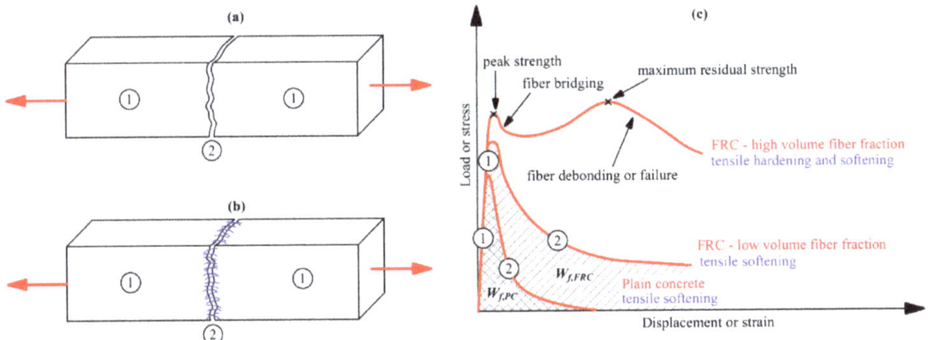

Figure 1. The behavior of plain concrete and fiber-reinforced concrete: (**a**) plain concrete; (**b**) fiber-reinforced concrete; (**c**) load–displacement diagrams.

The shape of the load–displacement diagram varies depending on the fiber volume fraction in the concrete, as well as the type of fibers used. There is a wide range of fibers on the market, which differ mainly in material, shape, shaped end, and the surface treatment of the fibers. Some of the types of fibers are shown in Figure 2.

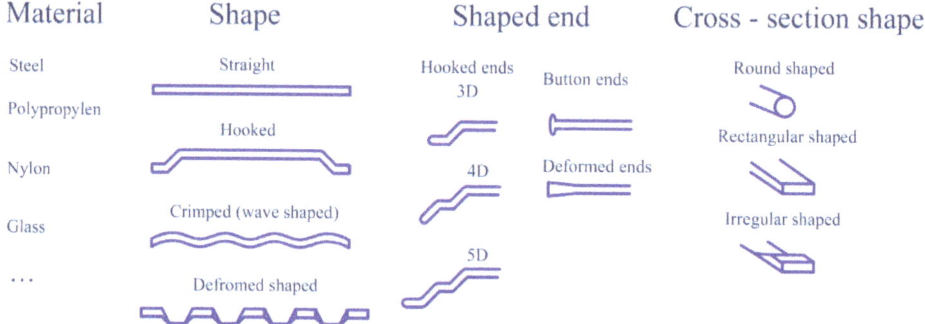

Figure 2. Shapes and types of fibers.

Most often, hooked fibers, straight fibers, or their combinations are used. The shape of the load–displacement diagrams ($f_c < 50$ MPa) for fibers with hooked fibers for different fiber dosages is shown in Figure 3 (blue lines). The figure shows the effect of the fiber volume fraction. By increasing the fiber volume fraction in the concrete, the absorbed fracture energy/fracture energy and tensile strength are increased.

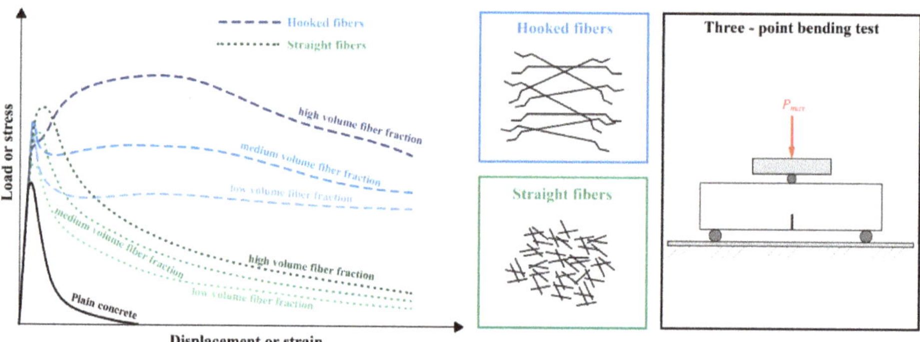

Figure 3. Load displacement diagram for fiber-reinforced concrete with hooked fibers and straight fibers [37].

Straight fibers are most often used to prevent shrinkage cracks. They can also be used in combination with other fibers that significantly increase the fracture energy and residual strength. In this case, this type of composite is called hybrid fiber-reinforced concrete. Figure 3 (green lines) shows load–displacement diagrams for typical fiber-reinforced concrete ($f_c < 50$ MPa) with straight fibers for different fiber dosages. The figure shows the effect of fibers on the tensile strength of the fiber-reinforced concrete and the effect on fracture energy. However, straight fibers are usually used for high-strength concrete, for which the load–displacement diagram may be more similar in shape to that of a load–displacement diagram for fiber-reinforced concrete with hooked fibers. For straight fibers, the quality of cohesion between the fiber and the concrete matrix is important, and in the case of high-strength concrete, this cohesion is significantly better. From a basic comparison of the load–displacement diagrams in Figure 3, it can be argued that hooked fibers usually have a greater influence on the tensile strength, residual tensile strength, and fracture energy increase. The fracture energy determined according to Formula (2) can be many times greater compared to that fiber-reinforced concrete with straight fibers.

3. Experimental Program

3.1. Recipe for Fiber Reinforcement Concrete

In the laboratory of the Faculty of Civil Engineering, VSB-Technical University of Ostrava, Czech Republic specimens with the following concrete recipe were tested: Min. cement content = 320 kg (CEM II/A-S 42.5); aggregate 0/2 = 525 kg; aggregate 0/4 = 420 kg; aggregate 4/8 = 150 kg; aggregate 8/16 = 820 kg; water = 200 l; and water/cement ratio (w/c) = 0.625. Parts of the laboratory program were samples that differed in the type of fibers. The hooked fibers Dramix® 3D 55/30 BG (Figure 4a) and straight fibers Dramix® OL13/.20 (Figure 4b) [38] with dosages of 40, 75, and 110 kg/m^3 were added to the concrete. The basic characteristics of the fibers specified by the producer Bekaert are shown in Table 1 [38]. For the classification of the concrete and the determination of the initial reference characteristics, test samples of plain concrete were also made, i.e., the fiber dosage was 0 kg/m^3. In total, seven test series were made (3× fiber-reinforced concrete with hooked fibers, 3× fiber-reinforced concrete with straight fibers, 1× plain concrete), which differed mainly in the type and dosage of fibers. Research on fiber-reinforced concrete followed.

(a)

(b)

Figure 4. Steel fibers: (**a**) Dramix® 3D 55/30 BG (hooked fibers); (**b**) Dramix® OL 13/.20 (straight fibers).

Table 1. General characteristics of fibers.

Material Properties	Dramix® 3D 55/30 BG	Dramix® OL 13/.20
Family		
Shape	Hooked	Straight
Bundling	Glued	Loose
Length l (mm)	30	13
Diameter d (mm)	0.55	0.2
Aspect ratio (l/d)	55	62
Tensile strength (N/mm^2)	1345	2750
Impact on concrete strength (kg/m^3)	25	60
Modulus of elasticity (GPa)	200	200

3.2. Determination of Mechanical Properties of Fiber-Reinforced Concrete

The mechanical properties of fiber-reinforced concrete/plain concrete were determined according to CSN EN 12390-3 [39], CSN EN 12390-5 [40], CSN EN 12390-6 [41], and CSN ISO 1920-10 [42]. The experimental program included the following examinations:

- The density, ρ: 3× 150 mm × 150 mm × 150 mm cube;
- 3-day compressive strength, $f_{c,cube,28}$: 3× 150 mm × 150 mm × 150 mm cube;
- 28-day compressive strength, $f_{c,cube,3}$: 3× 150 mm × 150 mm × 150 mm cube;
- Splitting tensile strength, perpendicular to the direction of filling, $f_{ct,sp,\perp}$: 3× 150 × 150 × 150 mm cube;

- Splitting tensile strength, parallel to the direction of filling, $f_{ct,sp,\parallel}$: 2× 150 mm × 150 mm × 150 mm cube;
- Bending tensile strength, $f_{ct,fl}$: 3× 150 mm × 150 mm × 600 mm beam with a notch in the middle of the span. The notch was 50 mm high and 3 mm wide. The load transfer device from the testing machine to the beam is composed of a support roller and one load roller. The support rollers were placed at a distance of 50 mm from the edge of the beam. The load was applied using a loading roller by displacement The load was applied using a loading roller by displacement increasing.

The test scheme is shown in Figure 5 according to [34,39–42]. Additionally, plain concrete tests were carried out for the static modulus of elasticity, E_c, for a cylindrical diameter of 150 mm and a height of 300 mm.

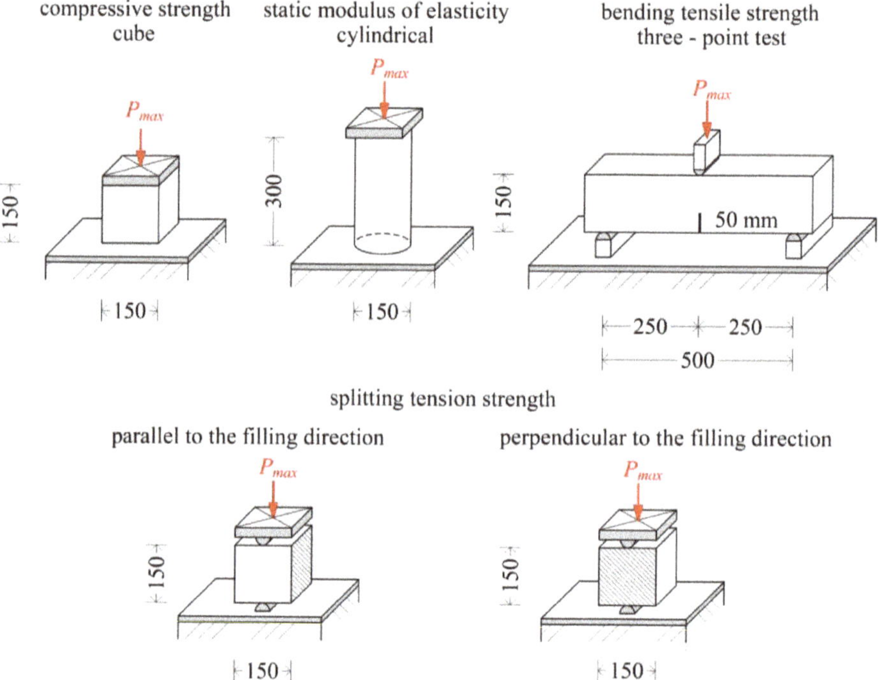

Figure 5. Test scheme in the experimental program according to [34,39–42].

The formulas used to calculate the basic mechanical properties were those recommended by Model Code 2010 [28]. For bending tensile strength, the formula is

$$f_{ct,ft} = \frac{3 \cdot P_{max} \cdot L}{2 \cdot b \cdot (h - a_0)^2}$$
(1)

where P_{max} is the maximum load achieved in the test press, L is the span, b and h are the transverse dimensions of the sample, and a_0 is the notch size.

The fracture energy G_f can be calculated from the load–displacement diagram and fracture plane. The fracture energy is defined by the relation

$$G_f = \frac{W_f}{b \cdot (h - a_0)}$$
(2)

where W_f is absorbed fracture energy, b is the width of the sample, h is the height of the sample, and a_0 is the depth of the notch.

4. Results of Laboratory Tests

The mechanical properties, derived from the laboratory experiment program, of fiber-reinforced concrete for individual dosages and individual types of fibers are summarized in Tables 2 and 3. Tables 2 and 3 also give the resulting mechanical properties of plain concrete. The average modulus of elasticity of plain concrete of 26.53 GPa was also determined. Two types of splitting tensile tests were performed to verify the homogeneity of the fiber-reinforced concrete, i.e., the even spatial distribution/orientation of the fibers in the concrete. This was a splitting tensile test perpendicular to the filling direction and parallel to the filling direction. The location of the test specimen in the press for the filling direction is shown in Figure 5.

Table 2. Material properties of plain concrete and fiber-reinforced concrete: compressive strength.

Type of Concrete	Dosing x (kg/m^3)	Average 28-Day Compressive Strength $f_{c,cube,28}$ (MPa)	Standard Deviation $\sigma_{fc,cube,28}$ (MPa)
Plain concrete	0	33.80	1.26
Dramix® 3D 55/30 BG	40	36.44	0.79
	75	40.92	1.64
	110	40.20	0.43
Dramix® OL 13/.20	40	38.45	2.02
	75	40.19	2.55
	110	42.83	2.32

Table 3. Material properties of plain concrete and fiber-reinforced concrete: tensile strength.

Type of Concrete	Dosing x (kg/m^3)	Average Splitting Tensile Strength $f_{ct,sp,\perp}$ (MPa)	Standard Deviation $\sigma_{fct,sp,\perp}$ (MPa)	Average Splitting Tensile Strength $f_{ct,sp,\parallel}$ (MPa)	Average Bending Tensile Strength $f_{ct,fl}$ (MPa)	Standard Deviation $\sigma_{fct,fl}$ (MPa)
Plain concrete	0	2.64	0.23	2.34 ± 0.01	3.90	0.15
Dramix® 3D 55/30 BG	40	2.81	0.20	2.41 ± 0.16	4.34	0.23
	75	3.97	0.29	2.70 ± 0.05	4.98	0.22
	110	4.75	0.47	2.91 ± 0.11	6.42	0.65
Dramix® OL 13/.20	40	3.07	0.23	2.70 ± 0.00	4.73	0.26
	75	3.08	0.40	2.51 ± 0.04	4.98	0.73
	110	3.08	0.51	2.59 ± 0.05	4.73	0.73

The average densities of the plain concrete and fiber-reinforced concrete are shown in the form of a bar graph in Figure 6. In each series (a total of seven test series), three samples were evaluated. Figure 6 also shows the standard deviations of the determined densities. The density was determined after 3 days of curing the samples stored in water. The initial density of the plain concrete was 2434 kg/m^3. In the case of the fiber-reinforced concrete, there was a gradual increase in densities with fiber dosing.

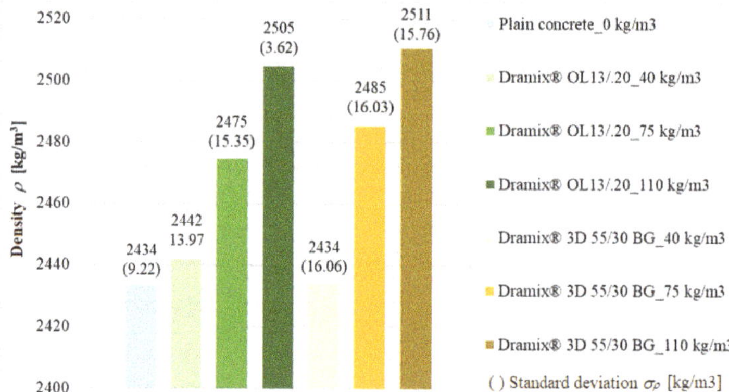

Figure 6. The average densities of plain concrete and fiber-reinforced concrete.

The 28-day compressive strength of the plain concrete was 33.80 MPa. There was an increase in compressive strength when increasing the dosage of the fibers. The increase in compressive strength compared with that of the plain concrete was up to 21% (fiber dosage of 75 kg/m^3) for fiber-reinforced concrete with hooked fibers and 27% (fiber dosage of 110 kg/m^3) for fiber-reinforced concrete with straight fibers. The values of the 28-day compressive strength are also specified in more detail in Table 2.

The initial increase in the compressive strength of the fiber-reinforced concrete was also observed. Specifically, the 3-day compressive strength of the fiber-reinforced concrete was monitored. These compressive strength values are shown in the form of a bar graph in Figure 7.

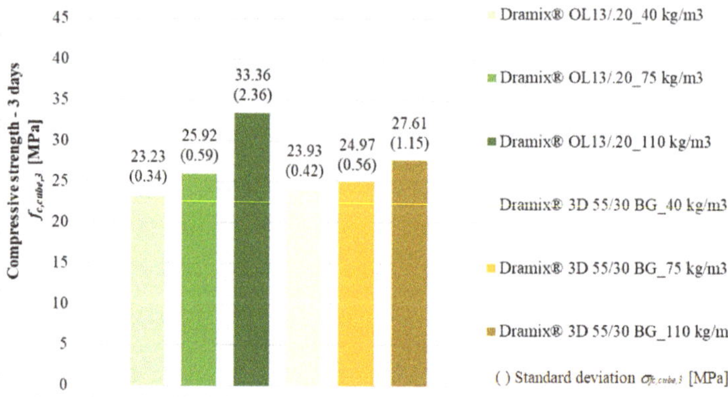

Figure 7. Three-day compressive strength of fiber-reinforced concrete.

Table 2 and Figure 7 show a difference between the average 3-day compressive strength and the average 28-day compressive strength of the fiber-reinforced concrete. In the case of the fiber-reinforced concrete with hooked fibers, the 3-day compressive strength was about 61–69% of the 28-day compressive strength, and for the fiber-reinforced concrete with straight fibers, it was 60–78% of the 28-day compressive strength. Figure 7 also shows, in parentheses, the standard deviations of the average values the 3-day compressive strength.

Table 3 shows the tensile strengths and their standard deviations. The increase in tensile strength was much greater for the fiber-reinforced concrete with hooked fibers. A comparison with the plain concrete revealed that the splitting tensile strength perpendicular to the filling direction increased from 2.64 to 4.75 MPa for the fiber-reinforced concrete with hooked fibers (fiber dosage of 110 kg/m^3), which is about 80%. The bending tensile strength increased from 3.90 to 6.42 MPa (fiber dosage of 110 kg/m^3), which is about 65%.

In this case, the fiber-reinforced concrete with straight fibers also had increased tensile strength, but the increase in strength was not as pronounced as that the fiber-reinforced concrete with hooked fibers. In comparison with the plain concrete, the splitting tensile strength perpendicular to the filling direction increased from 2.64 to 3.08 MPa for the fiber-reinforced concrete with straight fibers (fiber dosage of 110 kg/m^3), which is about 17%. The bending tensile strength increased from 3.90 to 4.98 MPa (fiber dosage of 75 kg/m^3), which is about 28%.

In comparison with the plain concrete, the splitting tensile strength parallel to the filling direction increased from 2.34 to 2.70 MPa for the fiber-reinforced concrete with straight fibers (fiber dosage of 40 kg/m^3), which is about 15%. In the case of the fiber-reinforced concrete with hooked fibers, the increase was from 2.34 to 2.91 MPa (fiber dosage of 110 kg/m^3), which is about 24%. The tensile strength (perpendicular and parallel to the filling direction) and the bending tensile strength of the fiber-reinforced concrete with straight fibers did not increase further with higher fiber dosages. In some cases, the tensile strength was even reduced with an increase in the fiber dosage.

The homogeneity of the concrete was verified by the ratio of the splitting tensile strength perpendicular and parallel to the filling direction. The splitting tensile strength ratios are shown by the bar graph in Figure 8. The graph shows a smaller difference between these strengths in the case of straight fibers, which is due to the better spatial orientation of these fibers.

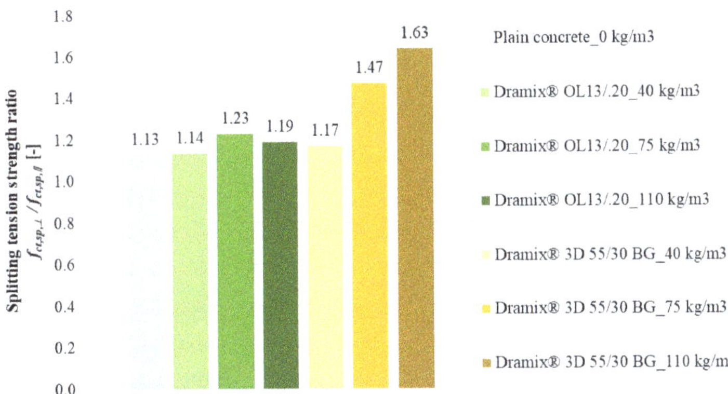

Figure 8. Splitting tensile strength ratio.

5. Evaluation of Mechanical Properties for Fiber-Reinforced Concrete

The data from the laboratory tests on the fiber-reinforced concrete with hooked fibers were evaluated using regression analysis. The functional dependence of the compressive strength (cube) $f_{c,cube}$, the splitting tensile strength perpendicular to the filling direction $f_{ct,sp,\perp}$, the splitting tensile strength parallel to the filling direction $f_{ct,sp,\parallel}$, and the bending tensile strength $f_{ct,fl}$ with respect to hooked fiber dosing can be expressed according to the following relations:

$$f_{c,cube} = 0.0695x + 33.8 \quad \left(R^2 = 0.8441\right) \tag{3}$$

$$f_{ct,sp,\perp} = 0.0175x + 2.64 \quad \left(R^2 = 0.8949\right) \tag{4}$$

$$f_{ct,sp,\parallel} = 0.0048x + 2.34 \quad \left(R^2 = 0.9164\right) \tag{5}$$

$$f_{ct,fl} = 0.0194x + 3.9 \quad \left(R^2 = 0.8902\right) \tag{6}$$

where x is the fiber dosage. On the basis of the coefficient of determination, R^2, the dependence of the mechanical properties of fiber-reinforced concrete on the fiber dosage, x, can be classified as very significant. Linear regression models fit the initial values of the mechanical properties of the plain concrete.

Graphically, the functional dependence of the compressive strength on the fiber dosing is shown in Figure 9. From the linear regression model, good agreement (84%) with the compressive strength data is evident. From Equation (3), it is possible to determine an estimate of the intermediate values of compressive strength. It is evident from the regression model that the increase in fiber dosing increased the compressive strength up to a fiber dosage of 75 kg/m^3. After this, there was a slight decline in this compressive strength (fiber dosage of 110 kg/m^3).

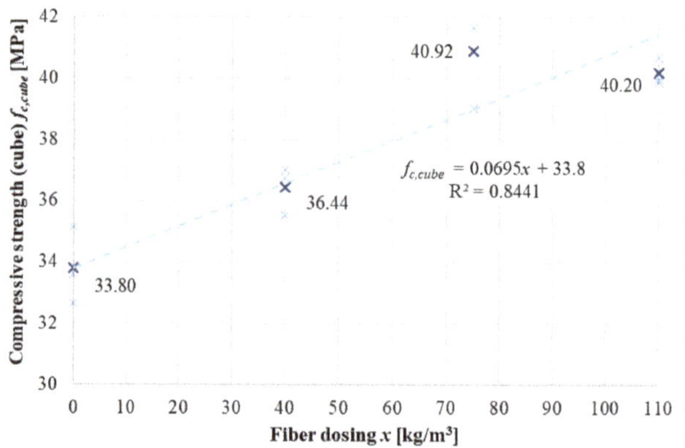

Figure 9. Regression analysis to determine the functional dependence of the compressive strength (cube) $f_{c,cube}$ on fiber dosage x.

Figure 10 shows the functional dependence of the splitting tensile strength perpendicular to the filling direction on the fiber dosage. Figure 10 shows a continuous increase in the splitting tensile strength perpendicular to the filling direction. The regression model shows good agreement, which is 89%. The estimation of the intermediate values of the splitting tensile strengths perpendicular to the filling direction can be performed using Equation (4). It is clear that the tensile strength perpendicular to the filling direction increased by up to 2.11 MPa.

Figure 11 shows the functional dependence of the splitting tensile strength parallel to the filling direction on the fiber dosage. Again, there was a gradual increase in this strength. This corresponds to the good agreement of the regression model, which is 92%. Intermediate values of the splitting tensile strength parallel to the filling direction can be estimated using Equation (5). Again, there was an increase in tensile strength, but it was not very significant.

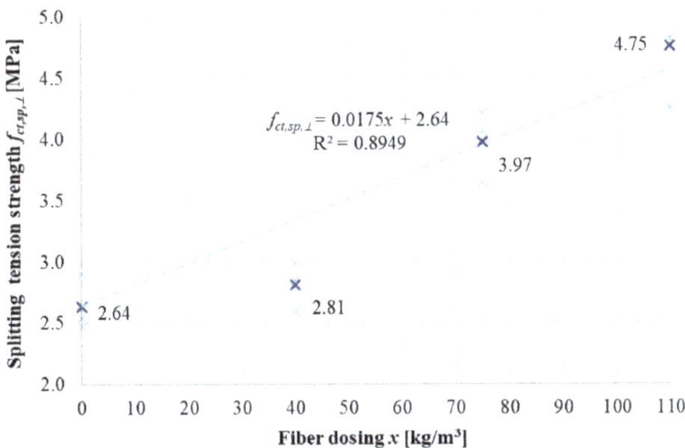

Figure 10. Regression analysis to determine the functional dependence of the splitting tensile strength perpendicular to the direction of filling $f_{ct,sp,\perp}$ on fiber dosage x.

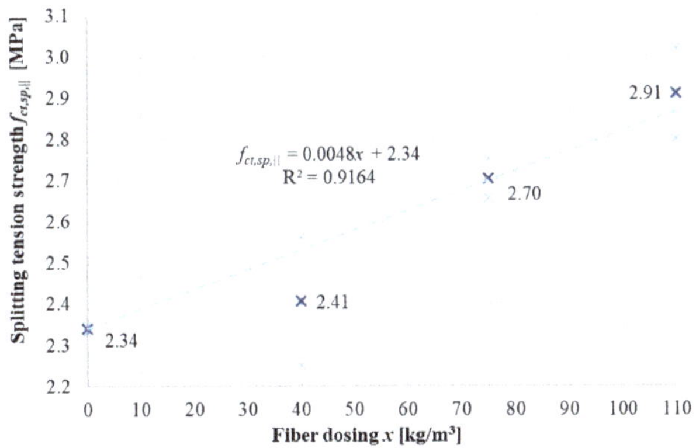

Figure 11. Regression analysis to determine the functional dependence of the splitting tensile strength parallel to the direction of filling $f_{ct,sp,\parallel}$ on fiber dosage x.

The functional dependence of the bending tensile strength on the fiber dosage is shown in Figure 12. It is possible to observe a gradual increase in strength with increased fiber dosages. Intermediate values of the bending tensile strength can be determined using Formula (6) with good agreement (about 92%). For all the tensile strength tests, the regression analysis was in good agreement and was around 90%.

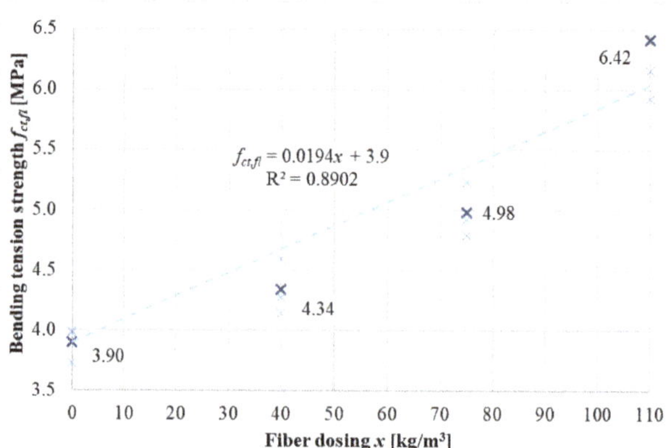

Figure 12. Regression analysis to determine the functional dependence of the bending tensile strength $f_{ct,fl}$ on fiber dosage x.

6. Comparison and Evaluation of Load–Displacement Diagrams

From the three-point bending test, load–displacement diagrams were evaluated. Figures 13–15 show the load–displacement diagrams for the plain concrete and fiber-reinforced concrete with hooked fibers and straight fibers, which differ in fiber dosage (40, 75, and 110 kg/m³). The black curves show the load–displacement diagrams of the plain concrete, the blue curves are those of the fiber-reinforced concrete with hooked fibers, and the green curves are those of the fiber-reinforced concrete with straight fibers.

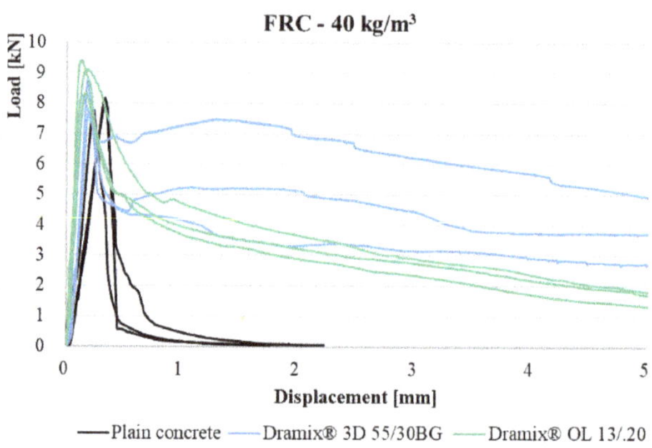

Figure 13. Load–displacement diagrams for plain concrete and fiber-reinforced concrete with fiber dosage of 40 kg/m³.

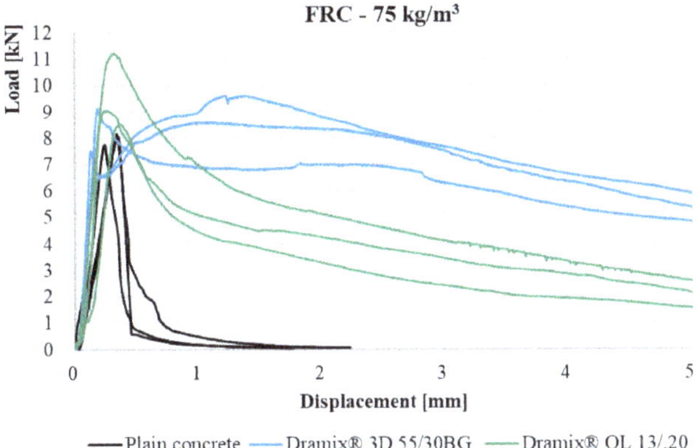

Figure 14. Load–displacement diagrams for plain concrete and fiber-reinforced concrete with a fiber dosage of 75 kg/m³.

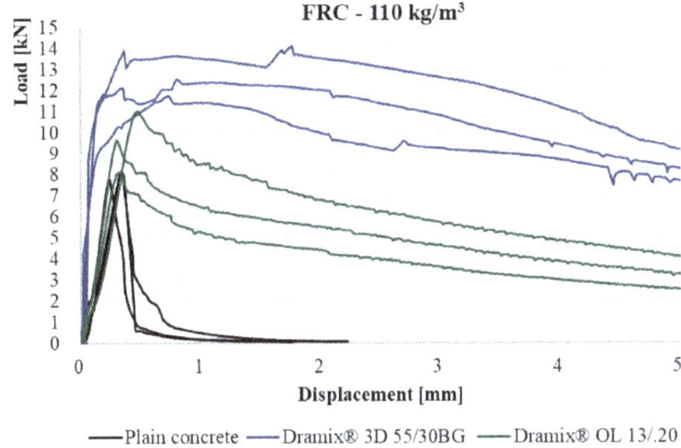

Figure 15. Load–displacement diagrams for plain concrete and fiber-reinforced concrete with fiber dosage of 110 kg/m³.

Figure 13 shows the curves of the load–displacement diagrams for the fiber-reinforced concrete with a fiber dosage of 40 kg/m³ and plain concrete. At first glance, it can be seen that the maximum load/peak strength at which the crack was located is slightly higher in the case of straight fibers. On the other hand, a different course of load–displacement curves for straight fibers and hooked fibers can be observed. The course of the load–displacement diagrams for the concrete with hooked fibers is more favorable. In the case of the fiber-reinforced concrete with straight fibers, tensile softening only occurs after the crossing of the point at which the crack was located (maximum load/peak strength). At first glance, the size of the absorbed fracture energy (area under the curve) involved in crack formation is also evident. In the case of the fiber-reinforced concrete with hooked fibers, this absorbed fracture energy is significantly greater. Furthermore, the residual strength is much higher for the fiber-reinforced concrete with hooked fibers. In the case of the fiber-reinforced concrete with hooked fibers, there is also tensile hardening and the formation of a second peak in the curve. This is the limit at which the fibers that bridge the crack were activated. However, the magnitude of the load of the second peak is smaller than the magnitude of the load at the first point, where the crack was located.

Figure 14 shows load–displacement diagrams for the fiber-reinforced concrete with a fiber dosage of 75 kg/m^3 and plain concrete. The course of the curve for the fiber-reinforced concrete with straight fibers is very similar to the course of the curve for the fiber dosage of 40 kg/m^3. Again, there is only tensile softening after the localization of the crack (first peak/peak strength). The course of the curve for the fiber-reinforced concrete with hooked fibers is very different. It can be observed that by increasing the fiber dosage, a more significant tensile hardening is induced. In two cases, the value of the load of the second peak on the curve exceeds the value of the load at the crack location (first peak).

Figure 15 shows load–displacement diagrams for the fiber-reinforced concrete with a fiber dosage of 110 kg/m^3 and plain concrete. The most significant difference can again be observed from the curves for the fiber-reinforced concrete with hooked fibers. The first peak, where the crack is located, is not very pronounced. There is a transition to the second peak without the presence of tensile softening. This course of the curve is also typical of high-strength concrete with straight fibers, which are added to the concrete to reduce shrinkage cracks.

Figures 16 and 17 show the load–displacement diagrams for the plain concrete and fiber-reinforced concrete with hooked fibers and straight fibers, which differ in fiber dosage (40, 75, and 110 kg/m^3).

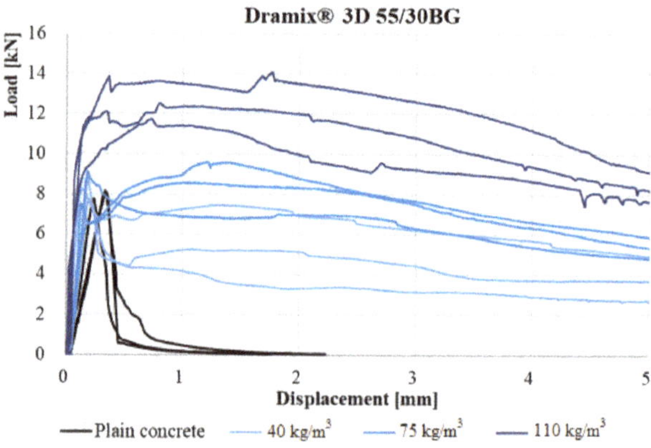

Figure 16. Load–displacement diagram for fiber-reinforced concrete with hooked fibers and plain concrete.

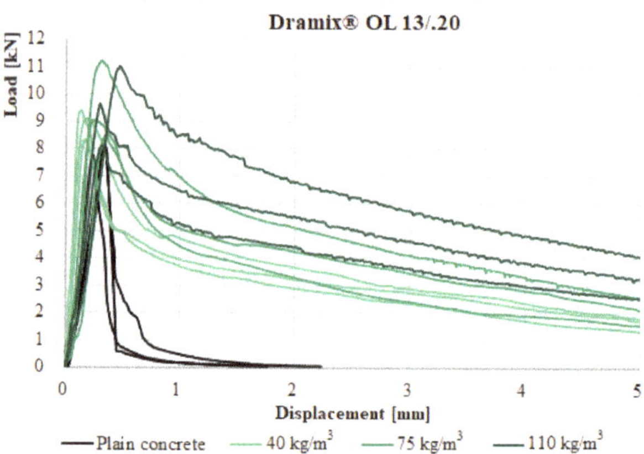

Figure 17. Load–displacement diagram for fiber-reinforced concrete with straight fibers and plain concrete.

It can be seen from Figures 16 and 17 that with an increasing fiber dosage, the tensile strength and residual tensile strength increase, but in comparison with fiber-reinforced concrete with hooked fibers, the increase is smaller for fiber-reinforced concrete with straight fibers. Figure 16 shows that with increasing fibre dosage, a significant change in the shape of the load–displacement diagrams can be observed. For the case of dosages of 40 and 75 kg/m^3, two sharp peaks occur (the first peak is crack location, and the second peak is maximum fiber activation). Between the start of loading and the first peak, only the concrete matrix cracks. After the crossing of the first peak, tensile softening occurs, when the fibers are still not fully activated. Tensile hardening continues to occur when the fibers are fully activated, which prevents further displacement until the second peak. After this peak, tensile softening occurs again, where the fibers break or are ripped out of the matrix. For the case of a dosage of 110 kg/m^3, the transition from the first peak to the second peak is smooth, and only tensile hardening occurs. The smooth transition is due to the greater dosage and cohesiveness between the fibers and the matrix. First, matrix cracks occur, and then, the fibers are rapidly activated.

Additionally, the effect of straight fibers was verified, as shown in Figure 17, which suggests that the fibers mainly prevent shrinkage cracks and do not have a big influence on the bearing capacity of the element/tensile strength. The residual strength of the fiber-reinforced concrete with straight fibers is small compared with that of the fiber-reinforced concrete with hooked fibers, and there is no second peak in the diagram. In this case, only tensile softening of the fiber-reinforced concrete occurs.

The average fracture energies corresponding to 5 mm displacement were also evaluated. The values are shown in Table 4. The table shows an increase in fracture energy with increasing fiber dosage in the concrete. For the fiber-reinforced concrete with hooked fibers, the increase is much more significant. When comparing the fracture energy of the fiber-reinforced concrete with hooked fibers, the increase between values of individual fiber dosages is an average of 49%. In the case of the fiber-reinforced concrete with straight fibers, the increase in the fracture energy between individual dosage values is an average of 23%.

Table 4. Fracture energy.

Type of Concrete	Dosing x (kg/m^3)	Fracture Energy G_f (N/m)
Plain concrete	0	152
Dramix® 3D 55/30 BG	40	1632
	75	2456
	110	3607
Dramix® OL 13/.20	40	1121
	75	1377
	110	1707

7. Determination of Shear Resistance of a Reinforced Concrete Beam without Shear Reinforcement

Based on laboratory tests for the structural verification of shear resistance, fiber-reinforced concrete with hooked fibers (Dramix® 3D 55/30 BG) was selected. The structural element was a reinforced concrete beam without shear reinforcement with dimensions of 190 mm × 100 mm × 1000 mm. This test was performed using a three-point bending test. The beam was provided with a reinforcement of 10 mm in diameter at the bottom surface. The coverage of the reinforcement was 20 mm. The test scheme of the beam is shown in Figure 18.

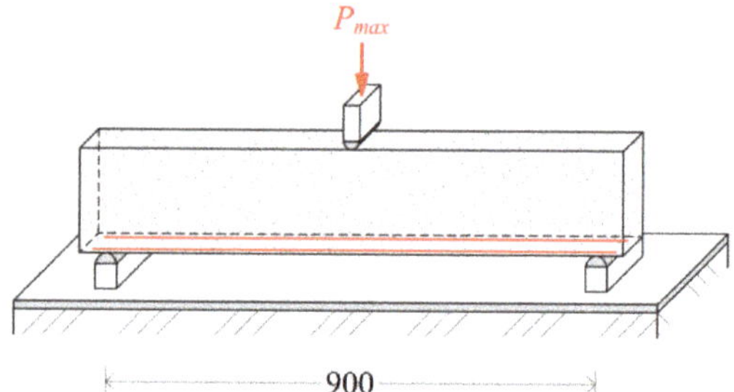

Figure 18. Test scheme of the beam with reinforcement.

Using regression analysis (Figure 19), the dependence of the maximum load P_{max} (at which the rays failed) on the fiber dosage was determined. It is possible to observe a gradual increase in the maximum load at which the failure occurred with increasing fiber dosage. Intermediate load values at which the failure occurred can be determined according to Equation (7). The model also shows very good data agreement, which is about 90%. The maximum achieved load for the plain concrete was 39.3 kN. Compared with the fiber-reinforced concrete, the maximum load increased by 47% (57.7 kN) for a fiber dosage of 40 kg/m³, by 64% (64.6 kN) for a fiber dosage of 75 kg/m³, and by 77% (69.6 kN) for a fiber dosage of 110 kg/m³. The functional dependence of the maximal load P_{max} on the fiber dosage x can be expressed according to the following relation:

$$P_{max} = 0.3088x + 39.3 \quad \left(R^2 = 0.8975\right) \tag{7}$$

where x is fiber dosage. On the basis of the coefficient of determination, R^2, the dependence of the resistance of the fiber-reinforced concrete (maximum load P_{max}) on the fiber dosage x can be considered very significant. Linear regression models fit the initial values (P_{max}) of the plain concrete.

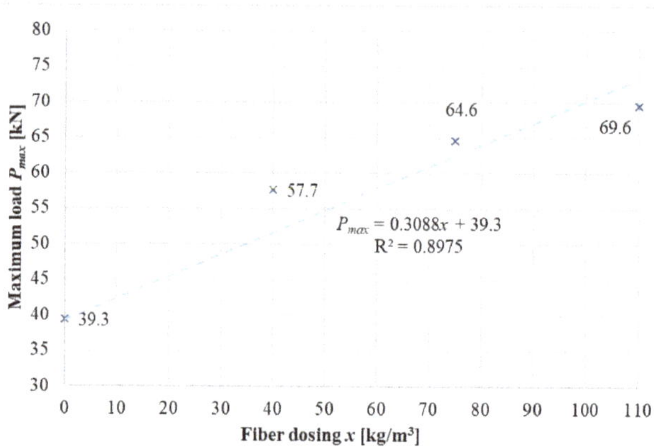

Figure 19. Regression analysis to determine the functional dependence of the maximum load P_{max} on fiber dosage x.

In Figure 20, it is possible to see the beams after a three-point bending test with a shear crack. The detail of the shear failure is shown in Figure 21. In all the fiber dosing cases, the nature of the failure was the same. It was a diagonal crack that widened from the lower edge to the upper edge of the beam during the three-point bending test.

Figure 20. Reinforced concrete (RC) Beams with shear crack.

Figure 21. Details of a shear crack.

8. Discussion

The positive effect of both types of fibers was confirmed by the evaluation of the compressive strength, which increased by up to 21% for the fiber-reinforced concrete with hooked fibers and up to 27% for the fiber-reinforced concrete with straight fibers. A more significant increase in compressive strength was observed for the fiber-reinforced concrete with straight fibers. On the other hand, a larger standard deviation for the compressive strength was found. As for straight fibers, it should be noted that their optimal use is in high-strength concrete, in which better cohesion between the fibers and concrete is ensured. Tensile tests allowed for more detailed comparisons. In the case of straight fibers, differences in the increase in splitting tensile strength and bending tensile strength were small for higher fiber dosages.

However, the standard deviation was greater for larger fiber dosages. The standard deviations of the compressive strength and tensile strength for the fiber-reinforced concrete with straight fibers were greater than those for the fiber-reinforced concrete with hooked fibers. Significant differences—an increase in the splitting tensile strength and bending tensile strength for higher fiber dosages—are evident for the fiber-reinforced concrete with hooked fibers. For all tensile tests, a regression function could be determined depending on the dosage, with a reliability minimum of 89%. For the 28-day compressive strength, the reliability of the regression function was 84%.

The verification of the homogeneity and spatial orientation of the fibers was an important area of this research. Splitting tensile tests were carried out for two variants: perpendicular to the filling direction and parallel to the filling direction. Control tests for the plain concrete showed the smallest differences between these tensile strengths. These small differences can be explained by the good compaction of the concrete layers. In the case of the fiber-reinforced concrete, the situation is further complicated:the homogeneity/fiber orientation must be taken into account when determining the splitting tensile strength. In the case of the splitting strength test perpendicular to the filling direction, the load was applied through the load roller to all the concrete layers that had formed in the samples as a result of the alternating placement of layers of concrete and compaction until the formwork was filled. When testing the splitting strength parallel to the filling direction, the load (load roller) during the test is situated in the place of the middle compacted layer of the concrete. Here, the influences of the size and shape of the fibers are introduced. In the case of the fiber-reinforced concrete with straight fibers, the maximum difference in the splitting tensile strength ratio was 1.2. The difference with plain concrete is small. Furthermore, a detailed visual study of the samples used confirmed that the fibers were spatially oriented in all directions. Greater differences were found for the fiber-reinforced concrete with hooked fibers. The maximum ratio of the splitting tensile strengths was as high as 1.6. With higher fiber dosages, the difference increased. A visual study of the damaged specimens also showed that when the fiber-reinforced concrete was compacted, fibers settled in the lower layers and were oriented horizontally. This particular problem can also be affected by the weight of the fibers.

Tests were also performed to determine the density. From the evaluated tests, a gradual increase in the density of the fiber-reinforced concrete relative to that of the plain concrete can be observed. The increase is due to the added fibers and also fiber dosing. The initial density of the plain concrete was 2434 kg/m^3. The density of the fiber-reinforced concrete is a very specific property, where large deviations in the determination can occur. The density is affected by many factors, especially the processing of the concrete and compaction. In the case of a higher fiber dosage, thorough compaction cannot be achieved because of the fibers. The rate of compaction also depends on the percentage of coarse aggregate contained in the concrete. In the case of a higher fiber dosage, it is more appropriate to modify the concrete recipe so that better workability of the concrete is achieved. The resulting density is also affected by the storage of the samples, i.e., the density of the natural storage, the density of the dried sample, the density of the sample stored in water, and the density of the fresh concrete mixture. The values of the stated the densities can vary by up to several percent.

A comparison of the 3-day compressive strength and 28-day compressive strength for fiber-reinforced concrete was also performed. In the case of the fiber-reinforced concrete with hooked fibers, the 3-day compressive strength was as high as 69% of the 28-day compressive strength, and for the fiber-reinforced concrete with straight fibers, it was as high as 78% of the 28-day compressive strength.

For a more detailed evaluation of the mechanical properties of fiber-reinforced concrete, bending tests are important. These tests also allow the post-peak behavior of fiber-reinforced concrete to be monitored. These are mainly residual tensile strength or fracture energy calculations. The resulting differences are very well distinguishable for different fiber dosages. In the case of straight fibers, the fracture energy increased significantly compared with the tensile strength. However, because of the poor cohesion of the concrete with the fibers, there was no tensile hardening. The load–displacement diagrams for the fiber-reinforced concrete with straight fibers were characterized by exponential curves. In the case of the fiber-reinforced concrete with hooked fibers and fiber dosages starting from 75 kg/m^3, there was also tensile hardening, and there were two peaks in the load–displacement diagram. For both types of fibers, all the specimens were compact after the test, even if they were considerably damaged (respectively deformed).

The shear resistance of the reinforced concrete beams with hooked fibers without shear reinforcement was also verified. Compared with plain concrete, there was a significant increase in shear resistance, and at a fiber dosage of 40 kg/m^3, there was an increase of 47%.

9. Conclusions

This article deals with the determination of the mechanical properties of two types of fiber-reinforced concrete with different fibers. These are hooked fibers and straight fibers. In total, three test series of samples were made for each type of fiber-reinforced concrete, which differed in fiber dosage (40, 75, and 110 kg/m^3). A series of plain concrete samples were also concreted and evaluated to determine the initial characteristics of the concrete. The laboratory tests also included the determination of the bending tensile strength and load–displacement diagrams. The splitting tensile strength (perpendicular to the filling direction) of the fiber-reinforced concrete with hooked fibers increased with an increase in the fiber dosage. In the case of the fiber-reinforced concrete with straight fibers, this strength did not increase at larger dosages; it remained unchanged. The uniformity of the distribution of the fibers in the test samples was also verified using the splitting tensile test. The uniformity of the fiber distribution was taken into account by the position of the sample, i.e., perpendicular and parallel to the filling direction. The splitting tensile strength perpendicular to the filling direction was, in some cases, as high as 1.5 times the splitting tensile strength parallel to the filling direction. The difference between these two strengths was larger with higher fiber dosages. The fiber-reinforced concrete with straight fibers allowed for the better spatial orientation of the fibers in the samples.

The results of the three-point bending test were also evaluated. From the load–displacement diagram, it is possible to determine the effect of different types of fibers (respectively dosage of fibers) on the resulting fracture energy and the shape of this diagram. In the case of hooked fibers, the increase in fracture energy was more pronounced: in some cases, it was as high as two times greater than that with straight fibers. The bending tensile strength increased with the dosage of hooked fibers. In the case of fiber-reinforced concrete with straight fibers, this strength did not show a significant increase.

The experimental testing also included a three-point bending test of a reinforced concrete beam without shear reinforcement and with hooked fibers to determine shear resistance. In the case of this test, the maximum load in the case of beam failure was monitored. By adding hooked fibers to the concrete, it was possible to increase the shear resistance by up to 77%.

The conclusions of the article are as follows:

- Reinforced concrete with straight fibers shows a better spatial orientation of the fibers in the concrete.
- The effect of fibers on the compressive strength is relatively small compared with that on the tensile strength.
- The density of fiber-reinforced concrete increases relative to that of plain concrete.
- To verify the homogeneity of the fiber-reinforced concrete, it is possible to use the splitting tensile test for the different directions of filling. The tensile strength is typically about 20% greater in the perpendicular direction to the filling direction than that in the parallel filling direction.
- Fibers in concrete always positively influence the tensile strength.
- For fiber-reinforced concrete with straight fibers, it is typical that with higher fiber dosages, the tensile strength increases very little, but the fracture energy significantly increases.
- For fiber-reinforced concrete with hooked fibers, it is typical that with higher fiber dosages, the tensile strength increases, and the fracture energy also significantly increases. Hooked fibers also have a significant positive effect on shear resistance.

Author Contributions: Conceptualization, R.C. and V.B.; methodology and validation, Z.M. and O.S.; data curation, Z.M and D.B.; writing—original draft preparation, Z.M.; visualization, Z.M. and O.S.; supervision, R.C.; project administration, Z.M. All authors have read and agreed to the published version of the manuscript.

Funding: This research was funded by VŠB-TUO by the Ministry of Education, Youth and Sports of the Czech Republic.

Acknowledgments: The work was supported by the Student Research Grant Competition of the Technical University of Ostrava under identification number SP2020/109. The writing of this paper was achieved with financial support by the means of the conceptual development of science, research, and innovation assigned to VŠB-TUO by the Ministry of Education, Youth and Sports of the Czech Republic.

References

1. Brandt, A.M. Fiber reinforced cement-based (FRC) composites after over 40 years of development in building and civil engineering. *Compos. Struct.* **2008**, *86*, 3–9. [CrossRef]
2. Lantsoght, E.O.L. Database of Shear Experiments on Steel Fiber Reinforced Concrete Beams without Stirrups. *Materials* **2019**, *12*, 917. [CrossRef]
3. Katzer, J.; Domski, J. Quality and mechanical properties of engineered steel fibres used as reinforcement for concrete. *Constr. Build. Mater.* **2012**, *34*, 243–248. [CrossRef]
4. Pająk, M.; Ponikiewski, T. Flexural behavior of self-compacting concrete reinforced with different types of steel fibers. *Constr. Build. Mater.* **2013**, *47*, 397–408. [CrossRef]
5. Abrishambaf, A.; Cunha, V.; Barros, J. The influence of fibre orientation on the post-cracking tensile behaviour of steel fibre reinforced self-compacting concrete. *Frattura ed Integrità Strutturale* **2015**, *31*, 38–53. [CrossRef]
6. Kohoutkova, A.; Broukalova, I. Optimization of Fibre Reinforced Concrete Structural Members. *Procedia Engineering* **2013**, *65*, 100–106. [CrossRef]
7. Sorelli, L.G.; Meda, A.; Plizzari, G.A. Steel fiber concrete slabs on ground: A structural matter. *ACI Struct. J.* **2006**, *103*, 551–558. [CrossRef]
8. Sucharda, O.; Bilek, V.; Smirakova, M.; Kubosek, J.; Cajka, R. Comparative Evaluation of Mechanical Properties of Fibre Reinforced Concrete and Approach to Modelling of Bearing Capacity Ground Slab. *Period. Polytech. Civ. Eng.* **2017**, *61*, 972–986. [CrossRef]
9. Cajka, R.; Marcalikova, Z.; Neuwirthova, Z.; Mynarcik, P. Testing of FRC foundation slab under eccentric load. In Proceedings of the fib Symposium 2019: Concrete—Innovations in Materials, Design and Structures, Kraków, Poland, 27–29 May 2019; pp. 380–386.
10. Zhao, J.; Liang, J.; Chu, L.; Shen, F. Experimental Study on Shear Behavior of Steel Fiber Reinforced Concrete Beams with High-Strength Reinforcement. *Materials* **2018**, *11*, 1682. [CrossRef] [PubMed]
11. Sahoo, D.R.; Maran, K.; Kumar, A. Effect of steel and synthetic fibers on shear strength of RC beams without shear stirrups. *Constr. Build. Mater.* **2015**, *83*, 150–158. [CrossRef]
12. Holschemacher, K.; Mueller, T.; Ribakov, Y. Effect of steel fibres on mechanical properties of high-strength concrete. *Mater. Des.* **2010**, *31*, 2604–2615. [CrossRef]
13. Karihaloo, B.L. *Fracture Mechanics of Structure Concrete*; Longman Scientific & Technical; Wiley: New York, NY, USA, 1995.
14. Löfgren, I.; Stang, H.; Olesen, J.F. Fracture properties of FRC determined through inverse analysis of wedge splitting and three-point bending tests. *J. Adv. Concr. Technol.* **2005**, *3*, 423–434. [CrossRef]
15. Abrishambaf, A.; Barros, J.A.O.; Cunha, V.M.C.F. Tensile stress–crack width law for steel fibre reinforced self-compacting concrete obtained from indirect (splitting) tensile tests. *Cem. Concr. Compos.* **2015**, *57*, 153–165. [CrossRef]
16. Giaccio, G.; Tobes, J.M.; Zerbino, R. Use of small beams to obtain design parameters of fibre reinforced concrete. *Cem. Concr. Compos.* **2008**, *30*, 297–306. [CrossRef]
17. Kormanikova, E.; Kotrasova, K. Elastic mechanical properties of fiber reinforced composite materials. *Chemicke Listy* **2011**, *105*, 17.
18. Karihaloo, B.L.; Wang, J. Mechanics of fibre-reinforced cementitious composites. *Comput. Struct.* **2000**, *76*, 19–34. [CrossRef]
19. Bakis, C.E.; Ripepi, M.J. Transverse mechanical properties of unidirectional hybrid fiber composites. In *Proceedings of the American Society for Composites—30th Technical Conference, ACS 2015*; DEStech Publications: East Lansing, MI, USA, 2015.
20. Koniki, S.; Ravi, P.D. A study on mechanical properties and stress-strain response of high strength concrete reinforced with polypropylene–polyester hybrid fibres. *Cement Wapno Beton* **2018**, *1*, 67–77.
21. Kasagani, H.; Rao, H.C.B.K. The influence of hybrid glass fibres addition on stress—Strain behaviour of concrete. *Cement Wapno Beton* **2016**, *5*, 361–372.
22. Lehner, P.; Konecny, P.; Ponikiewski, T. Comparison of material properties of scc concrete with steel fibres related to ingress of chlorides. *Crystals* **2020**, *10*, 220. [CrossRef]
23. Ponikiewski, T.; Katzer, J. Fresh mix characteristics of self-compacting concrete reinforced by fibre. *Period. Polytech. Civ. Eng.* **2017**, *61*, 226–231. [CrossRef]

24. Naaman, A.E.; Reinhardt, H.W. *High Performance Fiber Reinforced Cement Composites 2 (HPFRCC2): Proceedings of the Second International RILEM Workshop*; E & FN Spon: Ann Arbor, MI, USA, 1996.

25. Di Prisco, M.; Colombo, M.; Dozio, D. Fibre-reinforced concrete in fib Model Code 2010: Principles, models and test validation. *Struct. Concr.* **2013**, *14*, 342–361. [CrossRef]

26. Vandewalle, L.; Nemegeer, D.; Balazs, L.; di Prisco, M. Rilem TC 162-TDF: Recommendations of RILEM TC 162-TDF: Test and design methods for steel fibre reinforced concrete: Bending test. *Mater. Sci.* **2002**. [CrossRef]

27. *BS EN 14651:2005+A1:2007: Test Method for Metallic Fibre Concrete—Measuring the Flexural Tensile Strength (Limit of Proportionality (LOP), Residual)*; British Standards Institution: London, UK, 2005.

28. *Fib Bulletin 65: Model Code 2010, Final Draft–Volume 1*; Fib Postal: Lausanne, Switzerland, 2012; p. 350.

29. RILEM (2011): About Rilem [Online]. Available online: http://www.rilem.net/gene/main.php?base=50017 (accessed on 4 May 2011).

30. *DAfStbguidelines, 2011: DAfStb-Richtlinie Stahlfaserbeton*; Deutscher Ausschuss für Stahlbeton DAfStb: Berlin, German, 2011. (In German)

31. Köksal, F.; Şahin, Y.; Gencel, O.; Yiğit, I. Fracture energy-based optimisation of steel fibre reinforced concretes. *Eng. Fract. Mech.* **2013**, *107*, 29–37. [CrossRef]

32. Hoover, C.G.; Bazant, Z.P. Comprehensive concrete fracture tests: Size effects of Types 1 & 2, crack length effect and postpeak. *Eng. Fract. Mech.* **2013**, *110*, 281–289. [CrossRef]

33. Sucharda, O.; Pajak, M.; Ponikiewski, T.; Konecny, P. Identification of mechanical and fracture properties of self-compacting concrete beams with different types of steel fibres using inverse analysis. *Constr. Build. Mater.* **2017**, *138*, 263–275. [CrossRef]

34. Marcalikova, Z.; Prochazka, L.; Pesata, M.; Bohacova, J.; Cajka, R. Comparison of material properties of steel fiber reinforced concrete with two types of steel fiber. In *IOP Conference Series: Materials Science and Engineering*; IOP Publishing: Bristol, UK, 2019; Volume 549. [CrossRef]

35. Marcalikova, Z.; Prochazka, L.; Pesata, M.; Bilek, V.; Cajka, R. Mechanical properties of concrete with small fibre for numerical modelling. In *IOP Conference Series: Materials Science and Engineering*; IOP Publishing: Bristol, UK, 2019; Volume 596. [CrossRef]

36. Matsuo, S.; Matsuoka, S.; Masuda, A.; Yanagi, H. *A Study on Approximation Method of Tension Softening Curve of Steel Fiber Reinforced Concrete*; AEDIFICATIO Publishers: Freiburg, Germany, 1995.

37. Cajka, R.; Marcalikova, Z.; Kozielova, M.; Mateckova, P.; Sucharda, O. Experiments on Fiber Concrete Foundation Slabs in Interaction with the Subsoil. *Sustainability* **2020**, *12*, 3939. [CrossRef]

38. Bekaert. Available online: https://www.bekaert.com/en/products/construction/concrete-reinforcement (accessed on 28 April 2020).

39. CSN EN 12390-3: Testing Hardened Concrete—Part 3: Compressive Strength of Test Specimens. In *Czech Office for Standards, Metrology and Testing*; CEN: Brussels, Belgium, 2020.

40. CSN EN 12390-5: Testing Hardened Concrete—Part 5: Flexural Strength of Test Specimens. In *Czech Office for Standards, Metrology and Testing*; CEN: Brussels, Belgium, 2019.

41. CSN EN 12390-6: Testing Hardened Concrete—Part 6: Tensile Splitting Strength of Test Specimens. In *Czech Office for Standards, Metrology and Testing*; CEN: Brussels, Belgium, 2010.

42. CSN ISO 1920-10: Testing of Concrete—Part 10: Determination of Static Modulus of Elasticity in Compression. In *Czech Office for Standards, Metrology and Testing*; CEN: Brussels, Belgium, 2016.

MDPI
St. Alban-Anlage 66
4052 Basel
Switzerland
Tel. +41 61 683 77 34
Fax +41 61 302 89 18
www.mdpi.com

Crystals Editorial Office
E-mail: crystals@mdpi.com
www.mdpi.com/journal/crystals